Aussie STEM Stars

ALAN FINKEL

Australia's Chief Scientist: 2016-2020

Aussie STEM Stars

ALAN FINKEL
Australia's Chief Scientist: 2016-2020

Story told by KIM DOHERTY

Aussie STEM Stars series
Published by Wild Dingo Press
Melbourne, Australia
books@wilddingopress.com.au
wilddingopress.com.au

This work was first published by Wild Dingo Press 2021
Text copyright © Kim Doherty

The moral right of the author has been asserted.

Reproduction and communication for educational purposes:
The Australian *Copyright Act 1968* (the Act) allows a maximum of
one chapter or 10% of the pages of this work, whichever is the greater, to be
reproduced and/or communicated by an educational institution for its educational
purposes provided that the educational institution (or the body that administers it)
has given a remuneration to Copyright Agency under the Act.
For details of the Copyright Agency licence for educational institutions contact:

Copyright Agency
Level 11, 66 Goulburn Street
Sydney NSW 2000
Email: info@copyright.com.au

Reproduction and communication for other purposes:
Except as permitted under the *Copyright Act 1968*, no part of this book may be
reproduced, stored in a retrieval system, or transmitted in any form or by any
means without prior written permission.
All inquiries should be made to the Publisher, Wild Dingo Press.

Cover Design: Gisela Beer
Illustrations: Diana Silkina
Series Editor: Catherine Lewis
Printed in Australia

Doherty, Kim 1969-, author.
Alan Finkel: Australia's Chief Scientist: 2016-2020 / Kim Doherty

A catalogue record for this book is available from the National Library of Australia

ISBN: 9781925893434 (paperback)
ISBN: 9781925893441 (epdf)
ISBN: 9781925893458 (epub)

Success is the ability to go from one failure to another with no loss of enthusiasm.
   – **Winston Churchill**

Disclaimer
This work has been developed in collaboration with Dr Alan Finkel. The utmost care has been taken to respectfully portray, as accurately as memory allows, the events and the stories of all who appear in this work. The publishers assume no liability or responsibility for unintended inaccuracies but would be pleased to rectify at the earliest opportunity any omissions or errors brought to their notice.

Contents

1. Man on the moon — 1
2. Back on Earth — 11
3. A family mystery — 23
4. A school surprise — 31
5. Doctor who — 43
6. Planets and professors — 51
7. Opportunity knocks — 61
8. Out on his own — 70
9. Mountains of fun — 80
10. The unexpected call — 93
11. The chief challenge — 100
12. A brave new world — 112
13. High energy — 123
14. What's next? — 133

Afterword — 136

Glossary — 138

About Kim Doherty — 141

Dedication — 142

1

Man on the moon

'Finkel, Alan,' said the teacher, stifling a yawn as he monotoned his way through the Monday morning roll call from the front of the classroom. No answer. The class started to fidget. The teacher looked up from his clipboard. A few boys at the back nudged each other with knowing smirks. In the third row, Graham gazed innocently at the ceiling. Graham was one of Alan's best friends, and he was never innocent.

'Has anyone seen Alan?' followed up the teacher, suspiciously.

Alan was one of the best students in his year – possibly the whole school – but he wasn't exactly a goody-two-shoes. The teacher wouldn't have been surprised if Alan was too immersed in some new gadget design to realise class was starting. Or perhaps he was fiddling with the new circuit boards that had just been delivered to the science storeroom. But there was no sign of the lanky teenager anywhere.

And while his teacher looked across the desks and scratched his head, Alan Finkel was, almost literally, a world away...

The truant was curled up at home, on the carpet in his dad's den just off the kitchen, in front of their beautiful big television. The Finkels had been the talk of the street when it had been delivered. No one else in his area had a TV then, and all the kids had lined up on the front lawn to see it being unloaded. The glass screen was housed in a huge teak timber box, with cloth speakers and a smart silver dial you swivelled to move between stations.

But Alan wasn't going anywhere near that dial today. He wasn't even fidgeting, for a change. His bright blue eyes were staring intently at the black-and-white screen where, through a snowstorm of static, an astronaut in a bulky white spacesuit was stepping awkwardly down a ladder, about to change the history of the world.

*

Alan had been just eight years old when America's young President Kennedy had challenged his nation's scientists, at that time trailing the Soviet Union in the global 'space race', to put people on the moon.

'I believe that this nation should commit itself to achieving the goal, before this decade is out, of landing a man on the moon and returning him safely to Earth,' the president said.

Even to a little boy in Grade 2, far away in Australia, it seemed a crazy endeavour.

So, all through primary school, Alan had avidly followed every flight of the Mercury mission (one astronaut), every flight of the Gemini mission (two astronauts) and all the flights of the Apollo mission. This was before the internet, so Alan waited impatiently for each issue of his favourite magazines, *National Geographic* and *Scientific American*, to arrive in his mailbox, then he'd bury himself in the articles and the beautiful, glossy photographs.

Now, with President Kennedy's deadline only months away, Alan was about to find out if their best scientists, engineers and astronauts could work together to make that dream come true with Apollo 11. It was nail-biting stuff and he didn't want to miss a minute – even if it meant missing school.

But like any great adventure, things in space didn't go exactly to plan. It had all started well enough: Apollo 11 had blasted off on time from the Kennedy Space Centre in Florida, USA, and taken its scheduled 12 hours to get out of Earth's orbit.

Three days later, deep in space and hundreds of thousands of kilometres from their home, astronauts Neil Armstrong and Edwin 'Buzz' Aldrin had clambered out of the Command Module – leaving their crew mate Michael Collins on board alone – into the tiny lunar module, codenamed 'the Eagle'. They started their descent to the moon.

At that very moment, a world away, Alan was huddled in front of the television at home. There were no pictures yet – the moon was between the astronauts and the Earth, blocking the signal – but Alan could hear Armstrong and Aldrin talking from the Eagle to Mission Control in Houston, USA.

He pictured the men squashed in their insect-like craft, shooting through space, upside down, 10,000 metres above the moon and 384,400 kilometres from Earth – the farthest any human had ever been into space.

'Lights on. Down two and a half. Forward.' Armstrong's voice came steadily through the speakers. Seventy-five feet, looking good, down a half. Sixty seconds. Picking up some dust.'

Then suddenly, through the speakers, a warning alarm sounded. 'BEEP. BEEP.'

Alan jumped, he had never heard anything like it. Neither had the astronauts. It sounded again. And again. The Eagle's computer was straining, overloaded with data.

While Alan's heart thumped, Houston instructed Armstrong and Aldrin to ignore the insistent alerts and continue their mission.

The moon loomed larger through the Eagle's windows. Then the alarms sounded again. The computer, still flashing lights and sounding anxious alarms, was directing them to land in the crater below, but the men could see out the little windows that it was strewn with dangerous-looking boulders.

Armstrong took the controls. Overriding the computer and ignoring the urgent beeps and flashing lights, he calmly flew their little pod past the crater, looking out the window for a smoother place to land. The decision to fly on another six kilometres meant using more fuel. This created another problem: a new warning light flashed on – the pod had 60 seconds of fuel left.

Alan screwed up his eyes and listened, barely breathing, as Armstrong and Aldrin's voices bounced around the Finkel's den.

'Contact light.'

'Shutdown.'

'Okay. Engine stop.'

Their voices came through the speakers, steady but distorted through hundreds of thousands of kilometres of space. Then finally, loud and clear,

Armstrong's historic words: 'Houston, Tranquility Base here. The Eagle has landed.'

Alan imagined the Eagle, a leggy, mechanical intruder settling silently on the dusty, lifeless surface of the moon beneath a black sky. Inside it though, Armstrong and Aldrin were alive with anticipation – and much too excited to take the nap that someone back in mission control had written on their to-do list.

Instead, they set to work eagerly, resetting the systems so they could launch quickly in an emergency ('Asteroids? Aliens?' wondered Alan). Then they donned their suits for the long-awaited spacewalk. At last, Aldrin tugged at the door. It swung inward to allow Armstrong to step awkwardly outside onto the ladder. From there, he pulled a rope that released the TV camera mounted to their little ship.

The Eagle's camera beamed a black-and-white picture 384,400 kilometres from the moon to a remote tracking station in the Australian bush called Honeysuckle Creek. From there, the little Aussie base shot the images around the world.

> There were only four TV stations in Australia in 1969. The moon landing was broadcast on Channel 9, and it went for a record-breaking 163 hours, non-stop. Around the world, 650 million people watched the moon landing on television, thanks to the scientists at Parkes in NSW and Honeysuckle Creek in the ACT. And although none of them knew it, they had just changed Alan's life too.

Instantly, all over the globe, millions of televisions sprung to life, including one in a den in a suburban home in Melbourne, watched by an entranced teenager. Suddenly, there on the screen, Alan could see Armstrong jumping in weird slow motion down the Eagle's ladder and stepping, almost floating, onto the dust of the moon. The words he spoke at that moment have become legend.

'It's one small step for [a] man, one giant leap for mankind.'

Neil Armstrong and Buzz Aldrin changed the history of the world that day. And, although none of them knew it, they had just changed Alan's life too.

After his moon landing, Armstrong wrote to the Australians who worked at Honeysuckle Creek, Parkes and Tidbinbilla, the tracking stations that had played their part so well in the moon landing.

Down in Australia, there were some very dedicated people ... instrumental in the success of man's first flights to the moon. Science-fiction writers thought it would be possible ... to get people to the moon. But none [imagined you could] transmit moving pictures ... back to Earth. I was probably the most surprised person in the human race when Mission Control announced they were getting a picture ... It would be impossible to overstate the appreciation that we on the crew feel for your dedication and the quality of your work.

So I will just say 'thanks, mates',

Neil Armstrong

2

Back on Earth

The next day, Alan felt a little weightless himself as he bounded off the bus and handed in a sick note from his mum to the school office (throwing in a few well-timed sniffles as the principal's secretary looked at him sternly over her half-moon glasses) before heading to his locker.

'Fink!' he heard, a familiar voice shout. Across the milling kids, his oldest friend Graham and a bunch of his mates descended on him.

'Wagger!' exclaimed Joey, punching him playfully in the arm. Alan grinned at them.

'I don't know what you mean,' he said, coughing theatrically, then asking excitedly, 'Did you see it?'

'Only on the news last night,' said Graham.

The boys told Alan what he'd missed in class, then he told them what they'd missed on the moon.

Alan's friends knew, of course, that American astronauts Neil Armstrong and Buzz Aldrin had landed on the moon – you'd have to be living in another galaxy to miss it. But none of them had followed it with the fascinated attention to detail Alan had. When something interested him, Alan could be very, very focused.

'Not many jobs for a doctor in space,' laughed Graham over his shoulder, as they plunged into the throng of kids rushing through the corridors in all directions. Everyone knew Alan was going to be a doctor.

Alan chuckled on the outside. But on the inside, he hesitated. His friend was right. He'd thought the same thing himself.

That afternoon, Alan lugged his schoolbag home from the bus stop, carrying it over one shoulder like a sack filled with bricks. Most of them were science textbooks – he had some catching up to do. He dropped the lot inside the front door with a thud.

'Hello, Mum,' he yelled.

Their house was solid, double-brick, with strong, graceful arches and high ceilings. It was quite able to withstand the dropping of heavy bags from three kids, but it wasn't as sturdy as their previous home. Their Dad had built that one from scratch. In the little spare time he had from running his business, David Finkel had taught himself architecture and drawing skills, come up with a plan for the house and then worked with the builders to make a home for his family. Alan's dad was always learning something new, and he kept at it until he mastered it. It was one of the many lessons Alan was learning from his father.

'How was your day, darling,' his mum asked, indicating his afternoon tea, laid out ready for him on the kitchen table. She was wearing one of her favourite dresses, an orange floral with a wide skirt; it looked as if she was going out again. Vera Finkel was always being invited to some charity event or another; Alan tended to lose track.

'It was good,' Alan answered, munching. 'Can I go to Colin's?'

'Yes, but make sure you're home by 5.30.' She meant in time for dinner with Dad, their daily family ritual.

And just like that, with his mouth still full – Alan never sat still for long – he was off. Colin lived across the road and the two boys had been friends forever. They spent hours together, climbing trees in their cul de sac, or building billycarts, but usually they just hung out and read books. Colin wanted to be a doctor, too, and his family's set of medical encyclopaedias was better than any library Alan had seen. On that cold, wintry day, Colin was already sprawled on the carpet with the mammoth books scattered haphazardly around him when Alan arrived and flopped down on the carpet.

'Did you know,' Colin said without preamble, 'that the pancreas is here?'

He was poking at a Visible Man, a doll-size model of a human body, designed to teach students about anatomy. The skin could be peeled back and organs could be pulled out and stuck back in again. Alan had got it as a present and had lent it to Colin the week before.

'Um, yeah,' said Alan, watching Colin's pudgy fingers rearranging the Man's organs. He seemed to be putting the intestines back in upside down.

Alan and Colin were similar in many ways, but opposites when it came to body shape. Where Alan was tall and lanky, with aquiline features and a restless energy, Colin was heavily built and round in all directions.

He was much happier lounging about in the living room than dashing around outdoors. Colin once told Alan that he'd had polio as a child, which was why he had such a hefty build. Alan was never sure how true that was, but he let it go. Perhaps it was something they'd both learn about in medical school one day.

'Did you watch the landing?' Alan asked Colin. 'I wonder if the astronauts will survive the re-entry?'

Alan had read in one of his favourite science magazines that, in the coming days, the command module containing the three astronauts would be returning to Earth, travelling through the atmosphere at 15,000 kilometres per hour buried inside a fireball burning at over 2000 degrees Celsius. Then it would crash into the ocean and bob about in the waves until it was scooped up by the navy.

At least that was the plan. It seemed to Alan that the scientific calculations involved in getting the astronauts home safely were even more mind-boggling than getting them to the moon in the first place.

'Hmm, yeah,' said Colin. 'You know, I read somewhere that Armstrong didn't really plan on being an astronaut; he qualified as an engineer. Buzz Aldrin has a degree in engineering too, I think ... Ah, got it!' he smiled with satisfaction, turning the doll's intestines the right way up at last.

*

'How is your week going?' David Finkel's eyes twinkled at his youngest son across the dinner table. He knew exactly what Alan had been up to, but unlike most adults, he understood that it was curiosity and drive, not laziness or disobedience that had kept Alan away from school. His dad had a passion for learning, science and solving challenges too.

Alan shuffled his big feet under the table, and hastily changed the subject.

'My friend, Allan Lew, is thinking of specialising in cardiology when we finish school. I'm not sure, but perhaps I could be a surgeon? I like pulling things apart, fixing them and putting them back together,' he said, suddenly remembering the half-dismantled vacuum cleaner on his bedroom floor. He'd taken it apart, trying to understand what made it work. He'd better put that back together before somebody went looking for it.

'You'll be a wonderful doctor, whatever you choose,' his mum said encouragingly. His big sister Vivienne agreed. There were only two years between them, but he was the little brother she had cared for from the moment he arrived in the world. She knew him perhaps better than anyone.

'You've always been caring, even when you were a baby,' Vivienne said knowingly. 'And you still look after everyone, even, well, you know…'

Alan knew what she meant. He had a lot of close friends who got good grades and wanted to make their mark on the world like Colin, his chemistry lab partner Sam (Berkovic) and his good mate Allan (Lew). But he also had a few friends who walked to the beat of their own drum.

There was Leslie, who had been born with something not quite right. He was a goofy kid who would leave a trail of destruction behind him when he visited the Finkels, accidentally bumping into anything from the corner lamp to a vase on the table. But mostly, he was awkward with people, which made it hard for him to make friends and meant he got teased. But Alan would have him over to play, and make sure that Leslie felt liked and included.

Then there was Geoffrey at school, who was deaf, and for that reason was sometimes hard to understand. Alan would invite him home to hang out every now and then, and was friendly with him in the school grounds. Alan wasn't just good at his schoolwork, he was warm, inclusive and always fun to be around. No one would bully someone Alan called a friend.

'Oh, thanks,' he mumbled back to his sister, a little embarrassed.

He wondered if Vivienne was right. Perhaps he was caring. He supposed that would help him be a good doctor. He hoped so.

If you got good marks at school, it was expected by, well, everyone, that you'd go on to be a doctor or a lawyer. His big brother Ron was already studying law and commerce at university, but Alan knew the law wasn't for him. So that was it, he was going to be a doctor.

He enjoyed figuring out how the body worked and he loved fixing things that had broken around the house, so he was bound to like fixing people too. And it sounded good when he said it out loud to the chorus of approval from listening adults. He just had to ignore that little niggle of doubt that sprang into his mind every now and then. Alan wished it would go away.

*

Every dinner mattered in the Finkel household, but none more than Friday nights' Shabbat dinner when Alan's grandmother, aunts, uncles, his seven cousins and sometimes other friends came together at the table to eat, joke, chat and share their news.

> **Shabbat/Shabbos** is the Jewish day of rest and starts every Friday at sunset and ends Saturday sunset. It is customary for families to have dinner together on Friday nights.

Alternate weeks it was at their cousins' house, but Alan loved it best when it was at home – his mum was a gracious hostess and a brilliant cook. Vera always served at least three courses on a table that was beautifully set with elegant candles.

On special holidays, they had traditional Jewish fare, like challah bread and gefilte fish, but mostly their mum served up modern food. Dad would say the blessings over the wine, the bread and the family, which Alan always found a little curious. His dad had grown up in Poland in a deeply religious family – Alan's grandad had even trained as a rabbi – but amid all the awful things that had happened to him during World War II, his dad had lost his faith. So Shabbat dinner for the Finkels reflected their love for Jewish culture and this warm family tradition rather than the deeply religious event it was for many of Alan's friends.

'Shabbat is a time to give thanks for our blessings,' his dad would say. Alan knew his parents considered their blessings to be many. They'd come to Australia, worked hard and built a successful business and happy, safe life for their family. And unlike many Jewish families who had bravely made Melbourne their new home, the Finkels never felt they were victims of the horrors of the Holocaust. But of course, in some ways, they were.

The Holocaust

In 1933 the Nazi Party took power in Germany. The Nazis hated Jewish people and tried to make life hard for them. Later, during World War II (1939–45), they decided to kill as many Jews as possible. Their program became known as the Holocaust. It took the lives of about 6 million Jewish men, women, and children.

> Jewish people were not the only ones who died in the Holocaust. The Nazis also killed Roma (gypsies), homosexuals, mentally and physically disabled people, and anyone who dared to speak out against them.

3

A family mystery

Alan's mum grew up in Poland where her dad, her aunties and her granddad were famous milliners making beautiful hats for the men and women of Bielitz (now called, Bielsko) a town so close to the border with Germany that Vera and everyone she knew spoke German.

While Vera was happily going to school and synagogue and playing with her friends, her Mum and Dad were getting more and more worried. Just across the border in Germany, a group called the National Socialists – but usually shortened to the 'Nazis' – were growing stronger and louder. They blamed, demonised and dehumanised Jewish families for what they saw as the failings of the German state. Vera's parents feared war was coming, and they were no longer safe.

So when an aunt who had moved to Melbourne a few years earlier offered to sponsor the family to live in Australia, Vera's parents jumped at the chance. Alan's mum was 12 years old when she moved to the other side of the world and started her new life as an Aussie.

One year later, her parents' fears were realised, and Vera's childhood home of Bielitz was invaded by the Nazis. In coming years, her old neighbours were forced out of their homes, her school friends were pushed into ghettos and many of her extended family died in the Nazi death camps.

At the same time, on the other side of Poland, Vera's future husband was facing his own challenges. Alan's dad never told his children exactly what had happened to him during the war. It wasn't a secret so much as there always seemed to be something more interesting to talk about. But while he avoided talking about the war, his dad would happily recount his arrival in Australia in 1946, with his usual storytelling flair.

'When war finished, my brother and I managed to get tickets on the first steamship to leave Europe, a French one called the *Ville d'Amiens*, bound for Australia,' he'd tell his children with flourish. 'When I walked down the gangplank at Sydney's Circular Quay, I had absolutely nothing.'

Nothing in his pockets perhaps, but Alan knew he had bravery, initiative and kindness in spades.

'You had good looks and a large dose of charm,' his mum would say. She had fallen in love with him the moment they'd met, just after David arrived in Australia, and nothing much had changed.

'Well, I found someone who'd sell me some material, and I started my own business making clothes, just like my dad had taught me to do in Poland. From my first few sales, I started to save, and when I had a little money, you know what I bought? A new suit to impress your mum. I had it stitched in a special way so I could wear it inside out. Voila! I looked like a successful man with two suits!' he'd make them all laugh.

With an attitude like that, it was no surprise that after a few years of hard work, Alan's dad had built up a busy garment factory in Abbotsford, an inner suburb of Melbourne where there were lots of garment factories. He even designed the factory himself, poring over huge white sheets of paper night after night, drawing every detail – right down to designing the machinery and positioning the wiring.

Eventually his factory employed over 400 people, mostly migrants like himself hoping for a new start. And right from the beginning, without making a fuss about it, David would give 10 percent of everything he earned to charity. He never forgot how lucky he was to have survived the second world war that killed millions and to start his life in Australia with no money at all.

While amusing tales of his early days in Australia would often pop up at Shabbat dinners, it was his dad's life during the war that was a bit of a mystery to his children. Alan pieced bits and pieces together, mostly thanks to little slivers of information from chatty aunts and old family friends. They would mention life in Poland: a faraway place they loved, or someone's name from long ago, and then trail off…

'Your sister looks a lot like my sister Ruth did at her age...' they'd say, with a sad nod and mist in their eyes.

Eventually, Alan worked out the rough outline of the story, but by then he'd lost his chance to ask his dad himself.

It turns out that David grew up as one of seven children in Poland's buzzing, busy eastern city of Bialystok, loved for its warm summers, beautiful forests and thriving arts and culture. Bialystok was famous for making two things: textiles and vodka. The vodka makers found many willing customers across the nearby Russian border, while the textile makers like David's dad would sell rugs, carpets and clothes all over Poland and beyond. Half of Bialystok was Jewish in those days, and Alan's dad grew up speaking Yiddish at home and school, and Polish with the neighbourhood kids and, eventually, at work.

After he left school, David worked in his dad's successful business. His job was to cycle to villages to call on customers and take orders, particularly for their famous rugs. Over time, David gained a reputation for being bright, fun and trustworthy.

He built up a lot of friends in villages all over Poland. Perhaps this was why he wasn't home on the day it happened.

His family were together celebrating his younger brother's Bar Mitzvah, a Jewish coming of age tradition, when the Nazi's secret police – the Gestapo – banged on the door of their home and took two of his brothers away at gunpoint.

And perhaps it's why, when Bialystok was invaded by the Germans in 1939, David found himself with his mum, one brother and his sister on the side of the city where the Russians were in control – while his father and the rest of his family were caught on the German side.

The Soviets deported over a million Polish people to Russia's far east, bundling them onto trains and putting them to work in freezing factories, remote timber yards and terrible labour camps, to help with the war effort. That may be how David and his brother found themselves in deepest Siberia, working like they never had before, sharing the little food they had, trying to stay warm and not get sick. David looked after some of the children at the camp, particularly a

nephew he treated like a son. Quite simply, they did their best just to survive.

But as bad as life was in Siberia, things were even worse for David's dad, his other siblings, and his aunts and uncles back in the parts of Poland where the Germans had taken control. The family was forced out of their home by Nazis with guns, pushed into smaller and smaller ring-fenced sections of the city, called ghettos. Those who didn't die of cold or starvation or from being shot by soldiers in the ghettos, were sent to Nazi death camps. David never knew what befell most of his family. He never saw them again.

It was, Alan thought to himself when he discovered this, no wonder at all that his bright, positive, generous dad had always felt there was something better to talk about.

The underground movement in Poland

During World War II, almost half a million men and women joined the secret resistance movement in Poland against the Nazi invaders. They took incredible risks dynamiting rail transports, supplying the German military campaign on the Eastern front, making bombs and targeting roads and military targets, smuggling out intelligence reports to the Allies.

4

A school surprise

'Alan, time to get up,' he could hear his mum calling up the stairs. He was lying in bed, pondering his favourite poster pinned to the back of his bedroom door. It wasn't of an AFL football team – unlike most of his mates, footy had never really interested Alan – it was a photo of Earth, taken by Michael Collins, the Apollo 11 astronaut, while Armstrong and Aldrin were down on the moon.

'Our little planet looks so fragile, a swirling blue marble floating all alone in space,' he thought for about the thousandth time. It gave him an odd feeling inside. The two big global powers, the United States and the Soviet Union, were entrenched in a Cold War – a war of words and espionage. Every night the seven o'clock news reported the threats they threw at each other to launch nuclear weapons across the world. The military called it 'Mutually Assured Destruction' which was shortened to MAD.

'It sure is mad,' Alan muttered to himself. What was wrong with these grown-ups? Why would they risk this beautiful, fragile little planet we all call home?

> **Ham radio:** amateur radio using radio frequencies for non-commercial communication.

He shook his head, yawned and swung to his feet, neatly avoiding a mountain of books, a box of radio parts and a huge pile of science magazines – mostly *Electronics Australia*, *National Geographic* and *Science* – on the floor beside his bed. He'd signed up for a ham radio course that was advertised in one of them and the first box of parts had arrived the day before. Alan had stayed up late into the night trying to put it together, while his homework lay in his schoolbag untouched.

He was keen to test the radio out with Colin after school. He didn't really want to talk to anyone on the radio, he just wanted to see if he could make it work. The kit contained a tiny, busy circuit board, lots of wires and a sleek new plastic housing, so unlike the old wooden ones of his early childhood. He remembered the first time he saw plastic, tapping it and turning it over, trying to work out how it was constructed.

'Coming, Mum,' he answered, heading for the kitchen, pulling his school shirt over his head followed by his tie, which he hadn't bothered to untie from the day before. His dad had already left for work, and Ron for university, but Vivienne was at the table, starting on a plate of chops. Going to a Jewish school meant they didn't eat meat at lunchtime, so their mum made sure her kids ate a proper cooked meal at breakfast.

'Anything special happening at school today?' she enquired.

'Nope,' said Alan, without giving it much thought. It wasn't the first time he'd turn out to be wrong.

*

Later that morning, Alan and Sam were in one of the science labs, debating the instructions for their chemistry prac in loud whispers. As usual, Sam had conscientiously studied the pre-class reading and, as usual, Alan had scanned it quickly on his way to class. Despite this less-than-thorough approach to preparation, Alan won more of their debates than perhaps he should have, and the pair were neck-and-neck in their race to get the top marks in class. Mid-whisper, their teacher tapped Alan on the shoulder. Expecting to be in trouble, Alan looked up.

'Alan Finkel, you're wanted in the principal's office,' he said, waving him towards the door.

'Me?' Alan was confused. This didn't bode particularly well. He got good marks, held himself to high standards, and was kind and popular with both the kids and teachers, but he had in the past been described by the principal as 'a bit wild'. And the principal didn't know half of it.

The school's principal, Mr Ranoschy, was held in great respect and affection, not just by the students and teachers, but in the community.

He chose the best teachers, Jewish or not, and reminded the students it was their responsibility to be the best they can be, not just for themselves, but also for their community. He was determined that Mt Scopus School students would also be good citizens.

While he was well-loved, no one – not the teachers, the parents or the students – would ever consider calling Mr Ranoschy by his first name. It wouldn't have surprised the kids if he didn't have one. Harold reckoned that even when he was a baby, the principal's parents had probably called him Mr Ranoschy.

His name only added to the mystery that swirled around Mr Ranoschy. Like Alan's parents, the principal had been born in Poland before World War II. Rumour had it he'd survived the war as a young man by living and working on forged papers (to hide his Jewish origins) in the city of Krakow, while also secretly being active in the Polish Underground, part of the rebellion against the Nazis.

Alan had heard a rumour that Mr Ranoschy had once been called before the school board on a particular matter, where he'd stood up and told them in no uncertain terms, 'I wasn't scared of the Gestapo. Do you imagine I'm going to be scared by you?'

Alan couldn't imagine Mr Ranoschy being scared of anything. As he stood outside the principal's office, he wished he could say the same for himself.

'You can go in now,' said the receptionist. She was frowning at him over those glasses again.

'Alan,' Mr Ranoschy greeted him from behind his vast, carved wooden desk, 'please sit down'.

Alan nodded nervously, fiddling with his tie. He wished he'd tied it properly that morning.

'As you know, Alan, we had the election for school leaders a few weeks ago. I've calculated the votes and you've tied with another student, so the decision rests with me. I believe you have the academic ability, the support of your peers and the leadership skills to do great credit to the role, the school and yourself. I am, therefore, offering you the position of School Captain. Are you willing to take it on?'

Alan stuttered out an acceptance of sorts, 'Thank you, sir, I'll do my best'.

'Congratulations, young man, I'm sure you'll be a credit to us all. We will discuss your responsibilities in due course. Remember, leadership is a great honour, but it is also a great responsibility. Bear it well.'

'I will, sir,' said Alan.

That seemed to be the end of the unexpected conversation, so Alan stood to go. Then, treading across the carpet, he paused for a moment and turned back.

'Sir, can I ask you something?'

'I think you'll find you just have,' Mr Ranoschy smiled. 'But yes, of course.'

'What did you study?'

Mr Ranoschy looked a little surprised.

'Well, I have a Diploma of Education, and a Bachelor's degree in Arts and Engineering. I tried to study pure engineering in Poland, but my studies were interrupted by the war,' he answered. He looked at Alan thoughtfully. 'Is there something you want to discuss about your future, Mr Finkel?'

'No, sir,' answered Alan. 'I just wondered. Thank you.'

As he closed the door, he could feel his principal's eyes on him.

*

'School Captain?' Graham laughed when Alan quietly told his closest friends the news at lunchtime. 'I reckon I should tell Mr Ranoschy a story or two!'

They others smiled; they knew Alan was the right choice. They'd all voted for him themselves – but that didn't mean he was going to get off without a good ribbing.

'Does he know you dislocated my shoulder?' Graham prodded him playfully.

It was true. When they were at the local kindergarten together, Alan had grabbed one of Graham's arms from where he was standing on top of the monkey bars, and tried to haul his friend up strongly with both hands. Graham's shoulder had popped out of its socket. There was a rush of adults, and when Graham's mum arrived she whisked him off to hospital.

For months, four-year-old Alan had been terrified he'd done something terrible to his friend. He'd never been so relieved or delighted in his young life as when, at the beginning of the next year, Alan turned up for his first day at Mt Scopus School, only to see Graham standing there, too – with both his shoulders intact. It cemented a lifelong friendship.

'Or, how about we tell Mr Ranoschy you blew up the chemistry lab?' said Harold, joining in.

That was true too. Alan and Harold had been in the chemistry lab at lunchtime earlier that year, trying to work out how to make touch powder – an unstable chemical compound of ammonia and hydrochloric acid that exploded into purple powder at the lightest touch. They'd heard you could sprinkle touch powder on floors or lockers and it'd set off sparks if someone touched or stepped on it. The boys all thought it'd be hilarious if they could get the formula right for the upcoming muck-up day.

'I'm not sure it's working…' Alan had said impatiently, reaching out to touch their latest concoction with one finger. Before Harold could stop him… 'OOMPH!!' it had exploded all over the lab. Purple powder coated everything. It was all over the walls, the ceiling, and the boys. Alan and Harold had looked at each other in horror.

Just as they were contemplating the sort of trouble they'd be in, and wondering how on earth they'd clean all this up, their guardian angel walked in. Well, they didn't know she was their guardian angel just yet, but they would very soon. She was one of the school's laboratory assistants and instead of yelling at them or stalking out immediately to tell the head science teacher, she'd quietly just helped them clean it all up. And she never told anyone.

Perhaps Alan wouldn't be school captain now, if she hadn't been so kind and discreet that day. Alan didn't even know her name, but he was eternally grateful to her.

'Yeah, yeah,' he grinned at his friends.

He knew perfectly well that despite the stirring, they were delighted for him. And he knew his family would be too. The more he grew up, the more important it was to make them all proud. All of them, but especially his dad.

• • • • • • • • • • • • • • • • • • • •

Mutually Assured Destruction: MAD
When the United States finished World War II by dropping nuclear bombs on Japan, other countries realised how powerful nuclear weapons were – and many decided they wanted them too.

Within four years, the Soviet Union (a group of countries led by Russia) had developed nuclear weapons, and for the next 20 years, they raced the USA to have the most powerful weapons.

Each country threatened that if the other fired nuclear weapons, they would fire back, which would mean both countries – and pretty much the whole world – would be annihilated.

This was 'mutually assured destruction' (MAD) – the idea that both countries could destroy the other with the push of a button, so hopefully neither of them would actually do it. That would, indeed, be mad.

• •

5

Doctor who

Year 12 was always going to be a busy year, but with his new responsibilities, Alan found himself flat out every minute of the day, from his early mornings until he lay down to read every night. Apart from his studies, which tended to be a little haphazard and often rather last-minute, he was also juggling his commitments to cadets, his obligations to family and his new responsibilities as school captain.

There were assemblies to speak at, events to organise, projects to push. Alan, along with others, organised and led student walks for various charities, and lobbied the principal for a common room for the exclusive use of the Year 12 students, somewhere to relax away from the rowdiness of the younger kids. He had his School Captain's speech at Melbourne Town Hall to prepare for, and of course, there was that small matter of his final exams.

Alan knew he was lucky. He was born with a brain that just loved science and facts, solving problems and working things out. He was taking all his favourite subjects – maths, chemistry and particularly physics – all useful of course, if you're planning to be a doctor. His quick brain soaked up the information he read or heard, applied it comfortably and loved taking calculations to the next level – something that would be impressive at university level but was really exceptional in a high school student.

Alan wasn't sure where his talent for science had come from as neither of his parents were scientists although his dad seemed to have a natural flair for it. His sister told him she'd always known he was bright. She liked to recount the story of the day

Mr Potter had called. It had become part of the Finkel family folklore.

Vivienne had been in the kitchen with their mum one day when the phone had rung. It was one of those old-fashioned phones with a rotary dial, a curly cord attached to the wall, and a handset so big it was like putting a shoe to your ear.

'Good morning, Vivienne Finkel speaking,' she'd answered politely. The Finkel children had all been well-instructed in the etiquette of answering the family phone.

'Hello, young lady, may I speak to one of your parents, please? It's Mr Potter, Alan's teacher.'

She'd quickly handed over the phone to her mum, but she could hear Alan's Grade 5 teacher as his voice echoed out of the speaker and across the kitchen.

'Mrs Finkel,' Mr Potter had boomed, without much ado, 'are you aware that your son is a genius?'

Alan always rolled his eyes, embarrassed at the re-telling of this story. He suspected Vivienne had embellished it over the years, but he knew, too, that he was born lucky. And with that came expectations from his family, but also, from himself.

Having a good brain didn't mean though that he was always the most disciplined student, and with all Alan had going on, he'd found himself cramming more than he should. On the day of an exam, he'd get up super early – sometimes 4 o'clock – and jam a lot of information quickly into his brain. It would usually stay there long enough for him to get it down onto paper in the exam.

He was also lucky to have generous friends. Despite their good-natured rivalry, Sam had shared his brilliantly detailed chemistry notes with Alan, without which Alan knew he'd have been in trouble. It made him feel even more guilty when, eventually, he beat Sam at chemistry, 98 percent to 96 percent.

Year 12, too, brought decision time about their futures. In their limited spare time in their new common room, Alan and his friends would talk about their university ambitions and which course they were planning to list on their university application forms.

'Which medical course are you hoping for?' Colin asked him one day.

Their time together had been rarer this year.

They were both busy after school these days, and Colin was swotting particularly hard, trying his best to get the marks he needed to study medicine. He was determined to become a GP (a general practitioner), a local family doctor.

'Hmm, I'm not sure,' Alan answered vaguely. Melbourne University was traditionally the most prestigious, but Monash University was gaining a great reputation, too. 'I'm going to go to Monash's Open Day next week.'

'You know Sir John Monash was a German-Polish Jew,' said Colin, well aware of Alan's family heritage. 'He was supposed to be a brilliant engineer,' he added, almost as an afterthought.

Alan looked up at his friend. Colin's head remained firmly buried in his books.

*

It was only a few weeks later that decision day arrived. Alan was seated in the back row of the classroom next to Allan Lew, as their teacher handed out the all-important university preference forms. Every student in the room knew what they had to do: write down the courses they wanted to do after they left school in order: first choice, second choice, third. The universities would then consider each student's exam results and, if they were lucky, offer them a place in the course they wanted. Alan knew he was on track to get the marks he'd need for whatever he wanted to do – which, while fortunate, also made this decision all the more pressure-packed. This was the moment when they all had to commit to the career they wanted, perhaps for the rest of their lives. It was a big deal.

As the teacher got closer, moving through the rows, Alan could see his friends, one by one, taking the form, picking up a pen, and putting their heads down to write. Sitting at the desk in front of him, Alan knew Sam would be putting down Medicine – he was hoping to use his brilliance at chemistry to find a way to specialise in medical research.

Beside him, Allan Lew would carefully write down Medicine, too, with his hopes to be a **cardiologist**. Ever since they were kids, his

> A **Cardiologist** is a doctor who specialises in treating diseases of the heart and blood vessels.

friends had never wavered from their worthy and admirable ambitions. Alan wished he was as sure about his own career.

Instead, he just sat there, staring at the blank form, his thoughts churning.

'What are you waiting for?' Allan Lew whispered sideways at him, under his breath. 'You okay?'

'Um,' Alan took a deep breath, 'I'm not sure I want to be a doctor,' he whispered back. After all these years, faced with the sombre reality of the paper in front of him, it had finally hit him.

Allan Lew turned his head slowly, and looked at him. He didn't speak. Perhaps he couldn't.

'It's just that being a doctor means dealing with sick people. And old people. I like fixing things, but perhaps not people. I think it might be safer – especially for them – if I did something else...' Alan trailed off and buried his head in his hands. What was he doing?

But in his heart, he knew he was right. This realisation might have hit him at the last possible moment, but it wasn't too late to make a decision for his own future. This wasn't about his friends or his family. He knew they'd be surprised but they'd support him. This was his life. So, he took a big breath. Picked up a pen. And on the form in front of him he wrote a word that many admirable people – Neil Armstrong, Buzz Aldrin, Sir John Monash and Mr Ranoschy among them – had written before him: 'Engineering'.

> Not sure what exactly **engineering** is? Here's one way Alan explains it:
>
> Imagine a bridge designed by a scientist: of course, they would use the best possible materials, and consult the top experts. Great, right? Except it would be very expensive – so there's every chance the bridge would never be built.
>
> Now imagine a bridge, designed by an accountant, using as little money as possible. Sure, it saves a lot of money, but would you want to drive over it?
>
> Engineers find the middle ground. They know the science of perfect, and they know that money doesn't grow on trees, but nevertheless find ways to make it happen in a real, practical world.

6

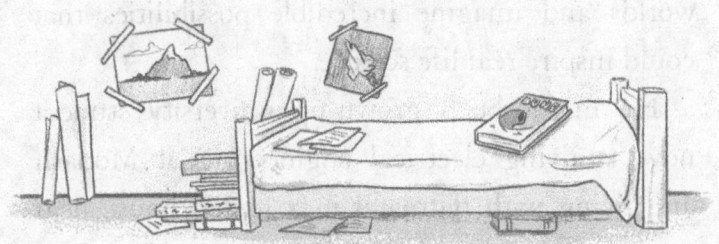

Planets and professors

Paul was having a few problems on the desert planet, Arrakis. He was trying to wrestle back control of the galactic empire and things weren't going too well. His mum had been sucked into quicksand and it was up to Paul to save her. Of course, Alan knew he would. He'd read *Dune* many times.

Alan had always loved science fiction books by authors like Arthur C. Clarke, Ray Bradbury, Ursula Le Guin, Jules Verne and Isaac Asimov.

He admired sci-fi authors who could create new worlds and imagine incredible possibilities that could inspire real life science.

He might be a grown-up university student now, studying electrical engineering at Monash and living with flatmates in a rented house near campus, but one thing didn't change: there was still a huge stack of books beside his bed. It was an ever-changing pile of science magazines, research papers, biographies and textbooks, but always near the top was his favourite sci-fi book – the 784-page intergalactic epic, *Dune*, written by Frank Herbert.

Set in the future, in a distant galaxy, *Dune* tells the story of Paul, a prince in a royal house who is sent from his comfortable home to the desert planet of Arrakis where he must learn quickly to survive (carnivorous sandworms and wild locals, for a start), so he can lead and protect his family in a fight for power in the galaxy.

As a teenager, Alan had simply loved Paul's adventures – weapons training, military strategy, learning to see the future and devising ways to rescue his mum from quicksand – but as he got older, Paul seemed to exemplify the sort of student Alan hoped to be. Over and over again, Paul would learn a skill from the best possible teachers and practise until he did everything well – then he was ready for whatever came his way. Alan wanted to do things well, too – not that he anticipated having to rescue his mum from quicksand anytime soon.

But as he lay in his lumpy student bed one night, thumbing through the well-worn pages again, it occurred to him that his beloved books gave him more than information and inspiration – they gave him a little break from the real world. And right now, that's what he needed most.

*

His dad wasn't well. Alan had noticed it at the end of his first year at university. His indefatigable father had survived Siberia and the war, worked hard building a huge business from nothing, and always had the energy to tell funny stories to his kids and host rabbis

> **Rabbi**
> A spiritual leader or religious teacher in the Jewish community.
>
> **Synagogue** or **Schule** (pronounced shool) is the place of worship for Jewish people.

to Shabbat dinner, and send them away with a generous donation for their synagogue. Now he seemed pale and tired.

The Finkel family had all ignored it at first, but Alan knew it was getting worse. Perhaps if he'd done medicine after all... He tortured himself with the thought. But he knew that a third-year medical student wasn't allowed to treat much more than the sniffles, anyway. And, even without a medical degree, Alan knew that whatever was ailing his dad was clearly much, much worse than that.

So instead, he brightened his dad's days with stories of university and the people he met at Monash who were just like the two of them. People who'd spend their days not just thinking about the theory of science, but how to take that academic knowledge to solve practical problems. How science and engineering could help astronauts go farther into space, or divers explore deeper under the ocean. How it might unlock the potential of people's brains or create robots that could make life better for people all over the world.

Engineering was a world Alan shared with his dad. His brother, Ron, had finished university where he'd been active in the Jewish student community – he'd even travelled to Israel and lived on a **kibbutz**. Now he was a successful lawyer and businessman. Vivienne was married and expecting a baby, and their mum was busy with the wider family and fundraising for all sorts of worthwhile causes. And they all did what they could to look after David; but he got sicker and weaker.

> A **Kibbutz** is a collective community in Israel traditionally based on agriculture.

As Alan spent more and more time with his dad, he found more common ground between them. And more admiration for the choices his dad had made and the person he'd become, despite all the horrors he'd lived through. It was his dad's work ethic, moral compass, generosity to everyone, passion for doing things well and positive attitude to life that Alan locked away in his heart. It was what he held onto when, later that year, his dad died.

Alan was only 21. Too young to lose his dad, but he knew that the best way he could honour his father and manage his grief was to focus on his studies, and to do things well.

So, everything Alan took on, he set himself the highest standard, and good marks followed. Then, as he approached the end of his degree, Alan found he had more decisions to make. In Year 12, he'd thought that when he finally decided on electrical engineering, that'd be it – he'd have a career path laid out before him like a sturdy ladder you could climb, step-by-step. But he soon discovered that there were almost as many specialties in engineering as stars in the galaxy. So here he was again, just like those last days in Year 12, wondering which specialty to choose, which ladder to step onto next.

Alan wandered around the Electrical Engineering Department at Monash University, talking to people about their work.

'I'm researching power systems engineering,' said one. 'You can come and try it out with me for a few weeks, if you like?' So he did.

'I'm doing material sciences,' said another. And Alan tried again, but it just didn't interest him enough to do it for years.

Then he met with an incredible lecturer and brilliant scientist called Dr Steve Redman, who was working in the department's Biophysics lab.

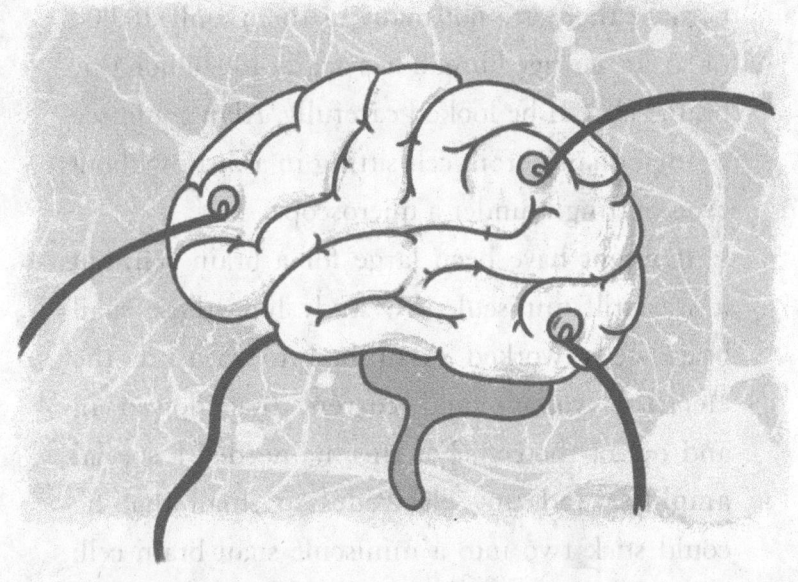

'We're researching electrical activity in brain cells. Are you interested?' Dr Redman asked. Alan smiled, the electrical engineering of a brain intrigued him. That's how he found himself studying neuroscience, how the brain works. And not just any brains ... Alan was about to spend the next six years of his life studying snails' brains.

You won't be surprised to hear that snails' brains are not very big. After all, slithering around a garden doesn't take a great deal of brainpower. But oddly, their brain cells are quite large – perhaps because there are not many of them, only 10,000 or so (an average human has nearly 100 billion tiny brain cells). If he looked carefully, Alan could see a single snail's brain cell sitting in a dish, without even putting it under a microscope.

It might have been large for a brain cell, but it was still miniscule. To study how those snail brain cells worked Alan had to measure the **electrical voltage** and **current** that flowed in and out of the cell. For this, he needed a special **amplifier** and tiny **electrodes** so small that he could stick two into a miniscule snail brain cell; one to measure the voltage and the other to send current in and out of the cell. He knew you could buy equipment like that in the United States but there weren't any in the Department of Electrical Engineering at Monash University.

Alan scratched his head.

He heard a noise behind him, and turned to see the Head of Department, Douglas Lampard, standing behind him. Alan knew the professor a little. He was a lovely man who often roamed the labs, encouraging students and asking about their work. But still, it was a little disconcerting to have him standing there, looking expectantly at Alan as he was scratching his head.

Douglas Lampard might be an encouraging teacher, an enthusiastic jazz musician and lover of great wine, but he was also a globally recognised scientist, a mathematician who'd had theorems named after him, and a very, very clever man.

Still, Alan figured, it was now or never.

'Hello, Professor,' Alan said, nervously. 'Um, I need a special amplifier, and a machine for making tiny electrodes small enough to get two of them inside a snail's brain cell. Would I be able to get these through the Department?'

Search these words and then check the Glossary section at the end of this book:

- Electric voltage
- Electric current
- Amplifier
- Electrodes

'Certainly, Alan. You may have any equipment you need...' said the professor in a friendly voice. Alan relaxed at the good news that a solution was at hand, '...as long as you design it and build it yourself'.

Alan's jaw dropped as Professor Lampard wandered off without another word. Alan's challenge had just turned into a much, much bigger one.

7

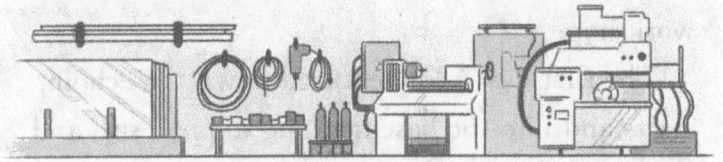

Opportunity knocks

Professor Lampard's words weren't empty ones. His students could have any equipment they needed – as he said, it could be the best in the world – they just had to make it themselves. But he didn't leave them entirely without help. In fact, he made sure they had everything they'd need.

There was a professionally equipped electronics workroom on the fifth floor of the university – staffed with brilliant technicians, many of whom would become Alan's friends – and, deep in the basement below, an extraordinary mechanical workshop.

The first day Alan ventured down the echoing stairs and into the basement, he stopped still and stared. It felt as if he'd wandered into a kind of wonderland. Stretched out across the huge, underground cavern were tools and machines of all shapes and sizes. They could whir, and cut, and shave, and beat metal into shapes that didn't even have names. Beside the machines were neat piles of metallic sheets, blocks of glimmering minerals, and silver and copper wires that glistened – all of it ready and waiting to be transformed into anything Alan could imagine. His dad would have loved this place.

On hand to help guide him through this underground domain was a gnarled old man named John, who was a wizard with it all. As a young man he'd been a Swiss clockmaker: the Swiss were famous the world over for the beautiful, precision work of their clocks, each one a finely tuned, beautiful creation that could keep time for generations.

John could use and fix every machine in his care, lovingly keeping them running and ready for the students to use in their latest creative quest. John knew the machines' faults and foibles and, best of all, could take a sheet of metal and transform it into something magical, all without saying a word.

Over the years to come, when Alan got stuck, or was dropping from tiredness from being up all night trying to get his latest design absolutely perfect, John would appear beside him, to suggest another way or simply give him a hand.

And on rare, special occasions, after many long frustrating nights of trying and failing, Alan would go home, dropping from tiredness. Returning to the basement the next evening he would find the tool he'd been clumsily labouring over for weeks waiting for him on the workbench – perfectly completed and gleaming under the fluorescent lights.

So, by day in the lab, Alan was learning from Steve, mastering the skills of a scientist: the research, the recording, the analysis. The importance of doing things well.

And by night, he'd be in the workshops, constructing fluid chambers, amplifiers and circuits, learning from John and the technicians so he could create and refine the tools he'd designed for his experiments.

He'd use his latest prototype tool in his experiments during the day. Then that night he'd take it back to his workbench, to tweak this or solder that, to make them even better, smaller or more accurate.

It was this tinkering time, in the evenings, sometimes through to 3 a.m., that Alan loved most. He was so focused, he'd hold off leaving his bench – even to go to the bathroom – until the last minute. And in the small hours after midnight, he'd finally fall asleep thinking about how he could make his latest design even better.

Then in the morning, he'd be back in the lab, with his tools, ready to test them out and learn more about snail's brains. Ready to do it all again.

Learn the principle. Do the practice. Apply the skills. Repeat. Just like Paul in *Dune*.

Most of all, he was determined to do everything well. By the time he finished his **PhD**, Alan was using the best electronic amplifier of its type in the world – which he'd invented. He just didn't know it yet.

But Steve did. He invited Alan to join him at the Australian National University in Canberra to do some new, breakthrough research. This time, the challenge was to do similar measurements but on much smaller cells, only 10 microns (one hundredth of a millimetre) in diameter.

The cells were too small for two electrodes to fit. Steve wanted Alan to make an amplifier that could use one electrode to do both tasks – measure voltage and pass current. For a scientist, it was like creating magic. Alan loved the challenge and worked and worked at it till, eventually, he had a solution.

∗

Then one day, Alan was hard at work when a famous American neuroscientist, visiting from New York, dropped by the laboratory and introduced himself.

'Good afternoon, Alan,' he smiled politely, reaching out a hand. 'My name is Paul Adams. I'd love to hear about your work.'

Alan gave him the usual spiel – the brain cells, the research, the results, the tools.

Looking over Alan's shoulder at one of his special amplifiers, Paul asked Alan a question that

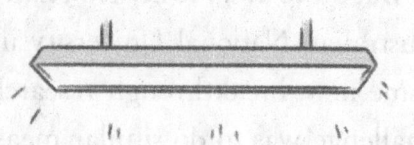

changed his life: 'Wow, Alan, could I get one of these?'

Alan looked at him for a moment as a light switched on in his head. Perhaps, finally, he'd become an expert in his field, and maybe this was the opportunity he'd been waiting for.

'Yes, Paul,' he said, slowly, 'I think you can'.

Later that night over dinner, Alan told his new girlfriend, Elizabeth, about his chance meeting. Elizabeth was strong, beautiful, creative, and very, very smart. She was a trained soprano who could sing like an angel. She was also pretty no-nonsense. They'd met at a friend's wedding when she'd noticed Alan watching her shyly from across the room. Elizabeth had walked up to him and asked him to dance.

'Everyone else might be intimidated by you,' she laughed later, 'but I'm not'.

Nor should she have been. Elizabeth was already a hugely successful scientist in her own right. She was studying for her own doctorate in **biochemistry** and had an impressive list of scientific work. Her brain was quite capable of challenging Alan's, and keeping him on his toes. Alan was smitten.

'I have some news for you, too,' Elizabeth told him over dessert. 'I've been offered a position at the University of California in San Francisco.'

It seemed like fate – San Francisco was a short drive to Silicon Valley, the global capital of electronics and the thriving home of the world's best digital technology. Alan didn't believe in godlike interventions, but it did seem to be a perfect opportunity.

'Wow,' he said. 'What better place to start a company making scientific amplifiers and software? Elizabeth, my love, I'll go with you.'

*

So, it was thanks to love, that Alan – now officially Dr Finkel, on his plane ticket – found himself boarding a plane bound for the USA. He had few concrete plans, little business experience and even less money, but he had his ideas, a willingness to work hard, and everything he'd been taught by his many inspiring teachers.

*

As he stepped out of the airport and into the Californian sunshine, after the 15-hour flight from Sydney, he thought of his dad who had also, many years before, arrived in a new land with nothing in his pockets. Alan reckoned he'd be smiling right now.

Masters of their domain

Paul from *Dune* might have been a fictional character, but there are real people, too, who have inspired Alan by being an expert in their field before they found their career path. For example:

1. Sundar Pichai, who was born in rural India. His family didn't own a telephone until he was 12 – so, of course, he didn't have an iPad, or a smartphone, or a laptop when he was growing up. His first degree was in **metallurgical engineering**. He went on to invent the Chrome web browser that you probably use every day. He became the chief executive of Google, but first, he was an expert in maths and physics.

2. Angela Merkel, who grew up in communist East Germany. At school, she did poorly at physics, but refused to be beaten by it and eventually earned a doctorate in **quantum physics.** She went on to become Chancellor (the leader) of Germany and the most powerful woman in the world. But first, she was an expert in physics.

8

Out on his own

While Elizabeth started her **postdoctoral research** work at the prestigious University of California in San Francisco, Alan bought a rattly old car and drove south to **Silicon Valley**, a region that was becoming world-famous for technology and innovation.

Silicon Valley isn't really a valley. Its wide, open spaces, fertile soil and sunny weather made it a kind of agricultural paradise and, until recently, there had been glorious orchards, teeming with fruit trees.

The laid-back towns, beautiful climate and proximity to top universities had attracted young, smart and energetic **entrepreneurs** to the area, keen to invent new products and launch companies. One company, called Intel, had invented microprocessors or 'chips' made of silicon – and a new nickname for the area sprang up.

Computers in those early days were enormous – the size of an entire room – they were clunky to use and so expensive that only governments, big businesses or universities could afford them. But gradually, these bright young people worked out ways to make computers small and user-friendly.

They dreamed up the ideas of screens that had graphics instead of numbers, and a 'mouse' to make things easier. Others thought up the ideas of 'clickable windows', 'icons' and 'menus'. Intel refined their chips, and in 1973 Xerox combined these ideas into a workstation, called the Alto.

> Why **'mouse'**?
> The first ones had cords attached to the computers, so inventors thought they looked like mice with a tail.

Computer clubs sprang up, a place where people could sit around together, have a drink, bounce ideas and share their technical knowledge. One night, at a computer club meeting, two young guys called Steve Jobs and Steve Wazniak brought along a home-made contraption they'd cobbled together in Jobs' parents' garage. It was the first Apple computer, which went on sale for $666.66 each (Wozniak liked repeating numbers!).

That's what Silicon Valley was like when Alan arrived. It was home to the world's fastest growing technology companies, and inhabited by smart, scientific people with ideas, energy and that American can-do attitude. Alan thought that if he couldn't start his business here, then he probably couldn't make a go of it anywhere.

So, he rented a spot in a little factory, set up a bench, ordered materials and rolled up his sleeves. With little money and no fanfare, Alan launched his company called Axon Instruments, named after the long, slender part of a nerve cell that carries electrical impulses.

In those early days, Axon was a one-person company so having a staff meeting generally meant Alan having a cup of coffee by himself. (He'd given up trying to find a decent cup of tea in America). It meant getting agreement on business decisions was pretty easy – but it got a bit lonely too.

Then one day, someone knocked on the door and stuck their head inside.

'I'm looking for Axon Instruments,' said the postman, looking curiously at Alan wearing his white lab coat and leaning over a workbench.

'You've found it,' said Alan with a smile, 'I'm Alan, the entire staff of Axon'.

'Howdy, Alan, good to meet you. I'm Eddie,' he said, stepping proudly inside to reveal the smart blue uniform of the US postal service and a small pile of mail. So began their daily exchanges.

At first Eddie would make an observation about the weather, or tell Alan what he'd seen on his rounds that day. Slowly their conversations moved on to family and home.

Then, one day, Eddie asked Alan what he was doing. Alan explained that he was making high-tech instruments to help scientists measure electrical activity in brain cells. He hoped his tools would help neuroscientists and researchers all over

the world find ways to treat diseases like **Parkinson's** and **Cystic Fibrosis**.

'Oh, Alan,' said Eddie excitedly, 'the people next door are doing the same thing'.

Alan jumped off his chair, ran to Eddie, put a hand on the old man's shoulder and looked him squarely in the eyes.

'Eddie, do you mean to say that I've come all this way around the world, spent all this time and the guys next door are doing exactly the same thing?!'

It turned out that Alan's neighbours were assembling machines to make the tiny electrodes that scientists would use with Alan's precision instruments. There was no other company in California doing that – just like there was no other company in California making instruments like Alan's, and there they were, next door to each other.

If Alan were spiritual, he would have said it was an omen. But as an engineer and scientist, Alan knew it was an amazing coincidence. What looked initially like a disaster, turned out to be a brilliant piece of luck, as the two companies started to work together. All thanks to Eddie, the postman.

Alan began to employ staff – carefully choosing people who were specialists in areas where he was not. And so he built a team who helped him create the best, most trusted and respected tools of their type in the world. But it wasn't all smooth sailing.

When he first worked out how much money it would take to use quality materials and employ great people, he realised he would have to sell his instruments at twice the price of his competitors. Uh, oh. He didn't have a lot of experience running a business, but it occurred to him that this might be a problem. In a panic, Alan rang home.

'Beep, beep,' went the long-distance line.

'Hello, Eric speaking,' came a voice.

Eric was his mum's new husband. Eight years after his dad had died, Alan's mum had remarried a lovely man who had become a wonderful support for all the Finkel kids.

'Hi, Eric, it's Alan ... um, I have a problem,' Alan said without much small talk.

International calls were expensive in those days. He told Eric about how much he'd have to charge for his tools in order to make a profit. Perhaps no one would buy them.

Eric listened, then asked, 'Alan, are your tools truly better than the competitors?'

'Yes, they are,' said Alan confidently. He wasn't sure of much, but he was certain about that.

'Then charge what you need to charge,' said Eric. 'Quality is remembered long after price is forgotten.'

Eric proved to be right, again and again and again. Alan sold his tools to universities, biotech researchers and medical companies that helped people all over the world. Everyone was asking to buy Alan's instruments. Then one day, many years later, someone asked to buy something else: his company. For $190 million dollars.

Alan called Elizabeth with the news. They'd married, and eventually they had both moved back to Australia to raise their two young sons, Victor and Alex.

But Alan still had his business in Silicon Valley to run, so for 17 years Alan had been flying between Australia and the United States. He'd spend time with his growing family in Melbourne, to be there to play with his boys, and tell them a science-fiction story before he put them to bed. Then he'd fly back to Silicon Valley, and back to work.

And so it went, at least ten times a year. Year after year.

He'd set up an Australian office of Axon Instruments, employing people just as energetic, smart and capable as those he'd found in America. The favourite bit of his job was to sit with the engineers and discuss the new features, and how they were developing them. He enjoyed challenging them to make the products better, just like he had been inspired by Steve Redman when he was a young student.

But it seemed with an offer like that, the time had come to handover and go home.

*

So, after landing in the US with nothing in his pockets 20 years earlier, Alan flew back to full-time life in Australia with millions of dollars in his bank account, and a plan to do, well, nothing much. Everyone said he'd worked hard enough, now he should relax and retire. But he wasn't too sure how that was going to go.

Elizabeth has a story she likes to tell about how Alan approaches problems. It goes like this:

'We'd been married a few years when we went on a holiday to Tasmania and visited Wineglass Bay. After walking to the edge of the bay, we took a nap on the sand. When we woke it was late afternoon, so we trekked back, but when we got to the car – uh-oh – no keys. They must have fallen out when we were napping.

It was an hour back to the bay but Alan jogged back, found our bed of rumpled sand, divided the area into a grid and methodically sifted through each square. Some nearby hikers got chatting to Alan, and next thing you know, they'd caught Alan's positive energy and wanted to help. Together, they found the keys in the very last sandy square!'

9

Mountains of fun

Alan's retirement was, in short, a disaster. He may have been older, but in many ways he was still the little boy who couldn't sit still. In the first few weeks, Elizabeth booked a relaxing family holiday in Byron Bay for them all. By breakfast on the first morning, Alan was restless. By lunchtime, Elizabeth had to go looking for him.

'Where's your dad?' She asked Victor, who was splashing around in the pool.

'I'm not sure – he's gone out. He said something about a pilot and a plane…'

Sure enough, Alan had found a company that offered light plane rides, and by the afternoon he and the boys were airborne, learning about the principles of flight, and how to fly a small plane. Right from the start, Elizabeth was pretty sure that retirement and Alan weren't going to get along.

Still, Alan tried. He reached out to his old friends. They might not have seen much of each other in the past 20 years while Alan was in Silicon Valley, but they were – and always would be – his best mates.

Allan Lew was living in Los Angeles in the US now – he'd achieved his boyhood dream of becoming a cardiologist. Sam was in medical research, making breakthroughs in the search for a cure for **epilepsy**. Colin had become a much-loved local GP. None of them had a lot of time to cavort around with their newly-retired friend. But Graham, David, another Alan and Joey were always there, ready to embark on Alan's latest madcap plan to learn to hang glide. And scuba dive. To go skiing. Alan got his motorbike licence. And a pilot's licence. And then he decided he wanted to climb enormous mountains.

So Alan took Elizabeth and his sons to climb Europe's highest mountain, Mt Blanc. Then he decided mountaineering was just the thing to do with his mates, too, to celebrate their milestone birthday year.

'What are we going to do for our fiftieth birthdays?' Alan had asked his friends. When they came up with vague answers about having a small party, or going on a holiday, he cut them off.

'Not this birthday,' he said, in his best school captain voice. 'You're coming with me. We're all going to climb Mt Kilimanjaro.'

And that's how Alan and his school friends celebrated their milestone birthday, lugging backpacks up Africa's highest mountain.

Mt Kilimanjaro, rising out of the African savannah, is one of the most striking mountains in the world. With Alan's long legs and happy energy leading the way, they all panted and stumbled their way upward, one step after another. Three long, exhausting and exhilarating days later, the group of old friends found themselves on the summit.

They sat together in the cold pre-dawn, surveying the incredible view as the sun's rays crept over the horizon, and shot out over the beautiful plains. It was a very special moment in their long friendship. David turned to Alan.

'This has been an amazing experience, Alan. Thank you *so* much for making it happen. But now, I think it's time you got a job. Before you kill us all.'

*

Alan knew David was right. Retirement wasn't for him. There was still so much to do – and, like his dad, so much he wanted to give back.

So, he embarked on some new projects.

Alan never forgot how, as a boy, he'd been informed and inspired by science magazines, so he and Elizabeth helped two science journalists, Wilson Da Silva and Kylie Ahern, launch a new science magazine called *Cosmos*. They'd met a few years earlier at a science dinner where Alan had fired questions at them at record speed. Wilson said Alan's brain was 'like a laser: sharp, sharp, sharp'.

Over a long lunch of pasta, they shook hands and launched the award-winning magazine for Australian kids. Alan wrote a column for the magazine called 'The Incurable Engineer', and eventually Elizabeth became the editor-in-chief. They'd tussle over his columns, with Elizabeth still challenging him to think more clearly, and write better.

These days, *Cosmos* magazine is posted to 87,000 Australian children each month (and 130,000 online), has won many awards and is probably in your school (or check it out at www.cosmosmagazine.com).

Alan asked some people he respected to be advisors to the magazine. Among the ones who accepted was Buzz Aldrin, the astronaut Alan had watched walk on the moon when he'd skipped school, all those years ago.

The chat with Buzz got Alan thinking. There was one science adventure he hadn't tried yet: going into space. He phoned Wilson late one night.

'Virgin Galactic are accepting bookings to take scientists and entrepreneurs into space,' Alan's excitement bubbled down the phone. 'Have you read about their new spaceship? What do you think of the technology? Is it safe? Will it work?'

Wilson knew Alan just wanted to chat to someone as passionate about space as he was.

'I love the idea of being up in that void, feeling the slow motion of zero gravity and seeing the black sky punctuated by stars that don't sparkle,' Alan continued, enthusiastically.

The two of them chewed it over: the technical difficulties, the enormous challenges, and the idea of Alan being part of science's greatest frontier.

'Would you want to go?' Alan asked suddenly. 'It'd be more fun together. And you could write about it and bring the adventure of space alive for the kids who read *Cosmos*.'

'Well, that'd be nice,' Wilson answered with a laugh, 'but I don't think I can afford it'.

Wilson later admitted he figured he had more chance of being named Secretary General of the

United Nations than going into space. So, he was possibly the most surprised person on Earth when Alan bought him a ticket.

As this book went to print, Virgin Galactic had just completed a test flight with a full crew on board – including the company's founder, Sir Richard Branson. They say their first passengers could be in space within a year. They've released photos of the spacesuits their passengers will wear, and the cabin with its seats specially designed to manage the G-forces, seat-back screens to display live flight data and 12 perfectly round windows. Each window has little handgrips you can hold when you're floating around the cabin so that passengers – including Alan and Wilson – will be able to look out together into the wonders of space.

*

In the meantime, Alan decided to become a director of organisations that helped make science more interesting for Aussie kids. There was the Australian Centre for Excellence of All Sky Astrophysics (which focuses on physics and space); a software company called Stile Education (which helps science teachers make science more interesting and interactive), and he became patron of the Australian Science Media Centre (to help journalists write better stories about science).

He even helped launch a new Australian university course in **neuroscience**, and took on the mammoth job of getting medical research institutes to work together as one institute. It's now called The Florey Neuroscience Institute. This means all the scientists working on different research areas on the brain can collaborate as a team and come up with more answers and solutions.

And, after staring at that poster of planet Earth every night of his childhood, Alan still believed science could help look after our planet. So he joined Better Place, a company that was trying to encourage more electric cars to be used across the world. And he started making lots of speeches all over Australia.

Then, just to confirm his retirement really was officially over, Alan was offered the position of Chancellor of his old university, Monash. His friend Martyn teased him, calling him the 'Grand Poobah' of the university.

On his first day back at Monash University, Alan wandered around the Engineering Department

where he'd spent so many happy years studying himself, stopping to chat to nervous young students about their work, just as Professor Lampard had done. And despite having a huge, important office now, Alan often wished he had more time to pop down to the workshop and tinker, just as he used to do.

*

Alan loved being busy and challenged, again, but he was never too busy to have adventures with his boys, dinner (and debates) with Elizabeth or regular Shabbat dinners with his mum, Ron, Vivienne and their growing families.

It was at one of those dinners that Alan told his family he'd been made an Officer of the Order of Australia (OA) for his 'distinguished service' to science and engineering, education, the protection of children and to philanthropy – helping the world. While his mum and Vivienne hugged him, and Elizabeth stood by proudly, Ron (who was soon to be made a Member of the Order of Australia himself) raised his glass to his little brother.

'After landing here with nothing in his pockets, Australia bestows one of its highest awards on his son. Dad would be proud,' Ron said, simply, knowing what it would mean to Alan. Alan smiled back, uncharacteristically lost for words. They both pretended they didn't have tears in their eyes. Little did either of them know that Alan's greatest honour was yet to come.

· ·

May the force be with you

When Alan first started making speeches, he wasn't sure how to finish them in a personal way. He wasn't really religious, so saying 'bless you' would feel a bit weird (especially as a scientist). But he'd always loved science-fiction stories, so the expression 'May the force be with you' from the Star Wars movies seemed to be a nice way to finish a speech and wish his audience positive energy in their life. Energy, positivity and science-fiction stories are all very Alan!

· ·

Great Aussie scientist

The Florey Institute was named after Dr Howard Florey was one of Australia's most famous scientists. He was born in Adelaide in 1898 into a family with four older sisters. His dad ran a boot-making company, but it was soon evident that 'Floss' – as the kids at school dubbed him – was set for bigger things. He was brilliant at science (although struggled a bit in maths), studied medicine at the University of Adelaide and in 1921 won a Rhodes Scholarship to Oxford University in the UK.

He was convinced that somewhere in nature there was an element that could destroy germs and, after many years of research at Oxford, he found a clue: an old paper by a Dr Fleming talking about the special properties of a kind of mould.

With his research partner, Dr Ernest Chain, he discovered how to extract the key ingredient of that raw mould juice and manufacture it into a medicine called penicillin – the first antibiotic – which has since saved countless lives. Fleming, Flores and Chain were awarded the Nobel Prize for Physiology or Medicine in 1945.

• •

Parable of the light cave

On one of their many family adventures, Alan, Elizabeth, Victor and Alex visited the famous Te Anau glowworm caves in New Zealand. They stepped onto a barge that glided silently through the water into the darkness of an underground cave and looked up to see thousands and thousands of tiny pin-points of light. It occurred to Alan then, that each of them was seeing that cave from a different point of view.

From her time as editor of Cosmos, Elizabeth knew a lot about the bioluminescence, the glow created by the nature world. So, what she appreciated was the science of nature, and incredible photos that might be good in the magazine.

Victor had become a businessman, so he appreciated how the Kiwis had created a business out of these caves. Alex, now a software engineer, appreciated how the tour operator kept improving the service for everyone who visited.

And Alan? Ever the incurable engineer, Alan saw incredible little creatures who could convert their dinner of mosquitos into light energy. They'd use their little lights to attract their next meal. Brilliant!

But the bigger lesson Alan took from the light cave was that whatever you have learnt, whatever you are an expert in, changes the way you see the world. It's part of what makes you, you.

• •

The first car to ever break 100 km/h was an electric car. It looked like a big bullet on wheels and broke the record in France in 1899. A year later, 38 percent of cars in America were electric, 40 percent ran on steam and 22 percent were running on other fuels like peanut oil. By 1915 there were no electric cars and no steam cars left – by then, the world had discovered cheap oil.

• •

10

The unexpected call

One day Alan was sitting at his desk when his phone rang.

'Hi Julie,' he answered, shuffling the papers in front of him. He'd been expecting her call. Julie was an employment consultant (sometimes called a headhunter, who finds the right person for the right job). Julie called from time to time, asking for Alan's suggestions: who'd be good at this job or that one. This time she was on a quest to find the next Chief Scientist of Australia. Alan knew some brilliant people who'd be wonderful at the job. He had his list of names ready.

'I have some great recommendations for you, Julie,' he said, 'do you have a pen?'

'Oh, Alan,' Julie laughed, 'I don't want your suggestions. I'm calling to talk to you about YOU. Everyone I've spoken to thinks you should be the next Chief Scientist. And so do I.'

'No, Julie,' he said, in a croaky voice that didn't sound quite like his own. No, no, no.

That night, he talked it over with Elizabeth.

'Do it, Alan!' she said. 'It feels as if you've been preparing for this role your entire life. And besides,' she added, 'now, more than ever, your country needs you'.

He asked his friends, and they said the same. Alan wasn't sure about taking on a job that was so public, so political and so different, but he still felt a calling to give back – just as his dad and Mr Ranoschy had taught him. And Elizabeth was right, there were so many challenges facing Australia. Perhaps he could help. Perhaps everything was falling into place, and this was the opportunity he'd been preparing for his entire life.

Perhaps he should just say yes. He looked up at his wife and smiled.

*

There were television cameras, bright lights, furry microphones and lots of people in smart suits busily checking their phones while, up on stage, the prime minister was speaking enthusiastically to the nation.

'Now more than ever, Australia needs to be more innovative, and more scientifically alert…' the PM waved his hands energetically in front of the television cameras, '…smart with technology, scientifically advanced, innovative, clever, imaginative…'

The Prime Minister turned to the side of the stage and smiled: 'That's why I'm so delighted that Alan Finkel has agreed to be Australia's new Chief Scientist'.

Everyone in the room burst into applause, and all eyes turned to Alan. He took a breath before stepping onto the stage and into the spotlight.

'Alan is remarkably qualified for this,' the PM continued. 'He is a scientist, an entrepreneur, an innovator and a communicator.'

Alan realised with a jolt, that this was all true.

He'd learned first to be a scientist, then he'd become a business entrepreneur with Axon, and was an innovator in almost every project he'd taken on. And through it all he'd communicated facts and ideas and enthusiasm, through his speeches and, of course, through *Cosmos*. It felt as if everything he'd learnt had led him here: learning from the best teachers, practising hard, and being brave and ready when the opportunity arose.

And this was the biggest opportunity of his life.

Alan stepped up to the microphone beside the Prime Minister and accepted this great honour, promising the Australian people he'd work hard for them.

'Although,' he thought to himself, 'I'm not exactly sure what my new job is'. And little did he know how big it was going to become.

*

Before Alan started his new job as Australia's 8th Chief Scientist, Alan and Elizabeth decided they'd celebrate with a holiday so Alan would come back refreshed and healthy and raring to go.

Alan had always wanted to see the Northern Lights – the beautiful-but-a-little-spooky coloured lights that sometimes appear in the sky over the North Pole. So off they flew to Iceland, Norway and Sweden.

'It's like being on the moon,' Alan said to Elizabeth, as he drove through the harsh, frozen landscape, dodging snow-topped boulders and large patches of ice on the road. Like the moon, this was a landscape that could kill you. They were out here in the Scandinavian countryside, miles from town, in the dark, where anything could happen…

It took many cold, scary drives on many dark nights, but finally they were in luck when the sky was clear and the conditions just right. Alan and Elizabeth looked up, entranced, as surreal sheets

of coloured lights danced across the sky. It was nature at its most beautiful.

And just to top off the experience, they'd booked to stay a night in the famous ice hotel which is made entirely of ice: the rooms were carved from ice, the tables and chairs were made of ice, even Alan and Elizabeth's bed was beautifully carved from ice in the shape of an enormous peacock. The wings stretched across the bedroom wall, the head and beak hovered over their bed as if he was a fairy-tale creature, watching over them as they slept.

The hotel had a chapel made of ice, an ice cinema and even an ice bar where Alan and Elizabeth toasted their incredible trip, and Alan's new job.

'Here's to your new job, your next adventure, my love,' Elizabeth said, raising her icy glass.

Alan smiled, his eyes were a bit watery. Perhaps it was the significance of the moment. Or perhaps not, because right then, Alan sneezed.

The next morning, Elizabeth called in the hotel nurse, who took one look at Alan and called the local hospital.

'I'm really fine,' Alan told the doctor weakly, as they bundled him off to the ward. 'I need to fly back to Australia and start my new job tomorrow.'

'Oh, I don't think so,' the doctor told him politely. 'You're off to bed. You have **pneumonia**.'

What causes the Northern Lights?

The sun doesn't just send light and heat out into our solar system, it also sends solar winds. These winds contain tiny protons and electrons. When a strong solar wind reaches Earth, it bumps into Earth's magnetic field (the one that makes our compasses point north) causing the protons and electrons to rush down towards our North and South Poles, until they crash into our atmosphere.

The collision of protons and neutrons with the gases in our atmosphere gives off a huge rush of energy as light, a different coloured light for each gas (oxygen gives off red and green light, nitrogen gives off blue). Over the North Pole, this incredible light display is called the Aurora Borealis; over the South Pole, it's the Aurora Australis.

11

The chief challenge

The day Alan walked into his new office in Canberra for the first time, he wasn't quite his usual hurricane of energy. He'd spent one night in hospital, flown home from Sweden and instead of going to bed, had jumped straight on a plane from Melbourne to Canberra, eager to start work the next day. The team thought their new boss seemed enthusiastic, energetic and ready to get things done. Little did they know how unwell he had been, or the huge amount of energy he usually had. It wouldn't take long for them to work it out.

*

When Alan had asked the Prime Minister, and the Minister for Industry, Science and Technology exactly what his new job was, it sounded simple enough:

1. Advise the government on science, technology and innovation and,
2. Be a champion for science, both in the community and internationally.

But Alan found out pretty quickly that what sounded simple was actually going to keep him very, very busy.

To start with, as Chief Scientist, he was the head of the Forum of Chief Scientists (also known as FACS) which included the Chief Scientist from every state in Australia and sometimes head scientists from other countries. They'd meet twice a year to discuss what was happening in science and research in their state – Alan could find himself talking about anything from Indigenous research to space exploration.

Then there was the National Science and Technology Council, a group that brings together science experts with the most important people in the government including the Prime Minister, to talk about science. What should Australia do about artificial intelligence? Or the internet of things? What about energy storage? Or synthetic biology? Or personalising medicine for everyone? Or how climate change will affect everything from what crops or fruit we will be able to farm in the future to the changes we need to make in designing our houses and cities. There were so many important topics to research and discuss, to make sure Australia was on the right track.

And Alan needed to know about all of them before he could offer the best advice to the government. Luckily, he loved reading; he was going to have to do a lot of it.

At the same time, the Minister for Science asked Alan to find out what big scientific equipment Australia's researchers would need for the next 10 years so they could keep doing world-class research. So Alan was off, meeting with people at Australia's biggest and most important scientific sites.

There was the *RV Investigator*, Australia's research ship that sails the ocean from Antarctica to our tropical north, studying everything from tiny krill to underwater volcanoes, while filming meteors and finding shipwrecks.

And the Square Kilometre Array (SKA), the world's biggest radio telescope which is being built in the desert of Western Australia. It will have thousands of dishes and up to a million antennas to allow astronomers to see more of the universe faster and in more detail than ever before.

Then there was the Australian Synchrotron in Melbourne. From the outside, the Synchrotron looks like a roofed football stadium, but instead of grass and seating inside, there's a maze of circular tunnels. Electrons are shot down these tunnels at about 299,792 kilometres a second (nearly the speed of light), producing light a million times brighter than the sun. This incredible glow allows researchers to look deep inside all sorts of atoms, from human cells to rare metals. Scientists use the discoveries from the synchrotron to improve people's health, look after the environment, and discover new materials and technology.

And while all that was going on, Alan's team was keen for Australians to get to know their new Chief Scientist. So Alan found himself doing interviews in all sorts of media, and giving speeches all over the country. It was just after one of these speeches that Alan was given his biggest challenge yet.

He had been asked to speak at Scotch College in Melbourne to mark the opening of their new science wing. As he was coming out of the new building, Alan saw Australia's Minister for Energy, Josh Frydenberg, across the crowd.

'Alan,' Josh waved at him above the melee of dignitaries, kids and parents and the sound of a marching band. 'Can we talk later?' he mouthed, holding up his hand in the shape of a phone. Alan nodded, waved back and promptly forgot all about it. A few of Alan's friends had come to see him speak and had then taken him out for a fun, rowdy dinner. They were all in the car, heading home at nearly midnight when Alan's phone rang. It was Josh.

'Alan,' said a sleepy voice at the other end, 'you know there have been blackouts in South Australia? Well, I've been speaking to the Prime

Minister, and we need someone to work out what's going on with Australia's electricity system and how to fix it. We think you're the man for the job.'

There was a yawn down the phone. 'I'm so tired ... G'night.'

There was a click, and Josh had gone before Alan had a chance to say a word. Alan sat there in the darkness of the car, staring at his friends. This, he knew, was going to be huge.

A massive storm had taken out the pylons and transmission lines in South Australia, leaving everyone without power for hours (in some places, days). A lot of people were blaming each other, and blaming the new solar and wind power for overloading the old system. It was Alan's job to sort out the facts, then work out what needed to happen to transform Australia's old electricity system into a safe, reliable world-class system that could cope with our needs now, and into the future.

*

'You'll be great at that!' said Elizabeth, sitting at the kitchen bench, trying to reassure Alan as he paced around the house later that night. 'You studied electrical engineering, remember?'

'Yes,' said Alan, 'I'm trained in electrical engineering, but my work has always been in microelectronics and electricity in a brain cell, which is so small you can't imagine how small it is. Power systems manage electricity flow that's so big, you can't imagine how big it is!'

Elizabeth continued on as if she hadn't heard him.

'And you went to a meeting in Adelaide last week to see if you could help with the blackouts.'

'Yeahhh…'

'You took the job of Chief Scientist to make things better in Australia, and this is a chance to do that, right?'

'Right,' Alan smiled at her.

Elizabeth smiled back. She knew Alan loved a challenge, particularly under pressure and that Josh and the PM were right: Alan was the right person for the job.

Sure enough, Alan was up at 4 o'clock the next morning, raring to go. His first call was to Gordon de Brouwer, the head of the government's Department of Energy. Gordon wasn't just smart and experienced – Alan trusted him.

Alan's first words were, 'Gordon, HELP!'

'Don't worry, Alan. Get over here and I'll get you started.'

And sure enough, they did. Gordon helped Alan put together a team of brilliant people, experts in the areas Alan needed, to look into what had gone wrong in South Australia.

It was like an enormous school project that took Alan nearly a year to finish. There were days when he'd wonder if he could ever get it done, but then he'd stare at a picture hanging at his desk, given to him by a friend. It was of a pelican, with a frog in its beak, about to swallow. You could just see the frog's arms reaching out of the beak, its hands wrapping around the pelican's throat, to stop himself from becoming lunch. Underneath, the caption said, 'Never, ever give up'.

That picture helped Alan keep going, especially on one particular day. He was in a **Senate hearing** being questioned by Malcolm Roberts, a senator from Queensland who had been a coalminer before he became a member of parliament in Canberra. He had a reputation for challenging science and liked to challenge Alan's ideas. Today was no exception.

'Isn't the scientist's first duty to be sceptical?' he asked Alan in front of everyone in the room, and many people watching at home on television. He was trying to set a trap.

Unworried, Alan answered clearly and politely.

'I think all the scientists I know have a healthy degree of scepticism. But *healthy* (and here, Alan leaned forward for emphasis) is an important word there. You have to have an open mind (Alan gestures at his ears) but not so much that your brain leaks out.'

For just a moment, Alan's eyes seemed to twinkle. Others in the room smiled. Alan had made his point.

In the end, everyone (except perhaps, Senator Roberts) thought Alan had done a brilliant job reviewing the electricity market – so much so, that it has become known all over the country as The Finkel Review.

That was when the government worked out what Alan's family had known since he'd taken apart the vacuum cleaner: he had a talent for fixing things. And everything he did, he did well. So it was no surprise that he didn't get to rest for long.

• •

As Chief Scientist Alan sometimes briefed members of the Department of Defence on new technology and science issues. On one memorable occasion, he was invited to spend a day underwater on a Collins Class submarine. He was thrilled when the Captain gave him the honour of holding the microphone to call out to the whole crew: 'Dive now, dive now'.

• •

Do you know what Alan's first decision as Chief Scientist was? He thought Australia's Chief Scientist should drive an energy efficient, kind-to-the-environment vehicle. So he went out and bought himself an electric car called a Tesla.

Elizabeth was doing a story for *Cosmos* on the whale sharks of Western Australia. Alan went along, diving off the boat with Australian scientist called Brad Norman to swim with the world's biggest fish. Brad had discovered that whale sharks each have a different spotty pattern on their backs, like humans have unique fingerprints.

Brad wondered if the pattern could help him track the whale sharks and find out why they were facing extinction. The trouble was, the pattern was difficult to record and recognise while the whale shark was swimming. Then a friend introduced him to a NASA scientist. Now Brad tracks individual whale sharks using the same software that NASA uses to identify star patterns in the sky.

What is synthetic biology?
'Imagine a future where synthetic jellyfish roam waterways looking for toxins to destroy, where eco-friendly plastics and fuels are harvested from vats of yeast, where viruses are programmed to be cancer killers, and electronic gadgets repair themselves like living organisms.' This is how *Cosmos* magazine explains synthetic biology, or 'synbio'. Scientists who specialise in synthetic biology view life as a machine – one that can be designed to solve all sorts of health, energy and environmental problems.

What is the Internet of Things?
The Internet of Things or IoT is the idea that any device could be connected to the internet and be able to talk to each other. If it has an on/off switch, that device could join the IoT. Which could be really convenient: imagine your alarm clock being able to tell your toaster to get started with your breakfast, or your car being able to tell your school if you're running late, or your fridge telling the supermarket you're running low on your favourite after-school snack. But, of course, there are worries too. Maybe someone could hack into your toaster and access your bank account! It's going to change our world, and we need to be talking about it.

12

A brave new world

Alan was starting to get the hang of his job, and the government worked out that Alan could help them solve all sorts of big challenges facing Australia. Before his five-year term finished, Alan did a whopping 12 reviews.

Every time, he gave the government the scientific advice they needed to make important decisions. But he was also still an engineer at heart – he wanted his advice to be practical. He knew that the solutions had to work in the real world.

'I think they should change my title to Chief Engineer,' he'd laugh to Elizabeth, 'or Chief Troubleshooter'.

He still got up at 4 o'clock, cramming hard for every speech, report or review. But there were three areas he worked on that were particularly close to his heart.

The first one was *education*.

Alan had always loved reading – wherever Alan was, a pile of books was never far away. As research shows, children who read a lot are more likely to succeed (so well done you for reading this!). Reading *anything* is good, from non-fiction books to romances, from science-fiction stories to tales of great adventure – it all helps.

So Alan set up the Story Time Pledge, asking grown-ups to promise to read to children. He also supported science teachers, started scholarship programs, spoke at lots of schools, encouraged everyone – especially girls, as we need more women in science in Australia – and launched the Star Portal, a website that helps families and teachers find fun science activities kids can do (You can too! Check it out at www.starportal.edu.au).

The second was *artificial intelligence* (AI).

Used well, computers, robotics and artificial intelligence can do wonderful things. Australian surgeons are now using robotics to perform delicate and complex operations through smaller incisions, sitting at a console with a high-definition, magnified 3-D view of the site of the procedure. Meanwhile, AI is improving the chance of healthy babies being born, preventing car crashes on our roads, creating beautiful art and bringing forgotten books to life.

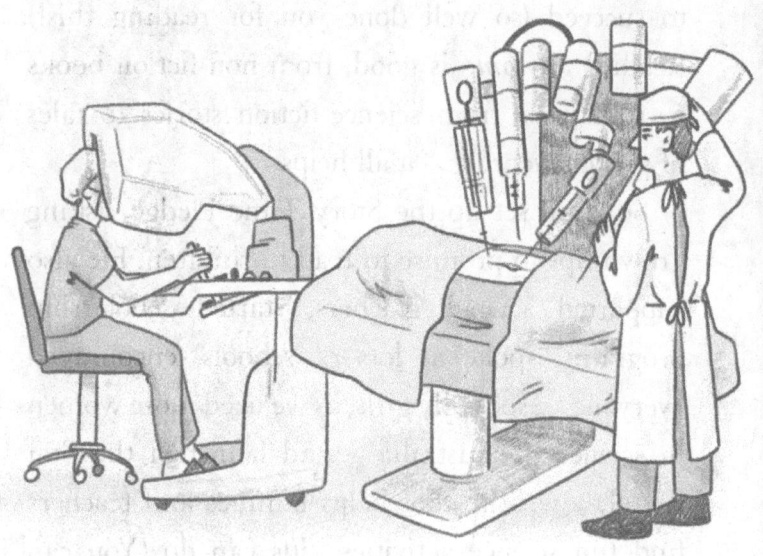

But used badly, they could do terrible, terrible things.

Many people in Alan's family had died in the Holocaust. The Nazis used a very efficient data system to identify millions of people to send to the death camps. Their system was a sort of precursor to computers. If the Nazis had had the power of today's computers, the terrible things they did might have been even more horrific.

Alan was determined to talk about what sort of world we wanted to live in, and make sure that robots and artificial intelligence helped build a better world, not a worse one.

He proposed a few rules about robots and humans living together that might help. For example, that humans should always be the boss of robots. That, as robots get cleverer, so should humans – we need to keep learning. And that people who develop and control robots mustn't be greedy; they have a responsibility to make a better world for everyone.

But perhaps Alan's biggest passion was *taking care of our planet*.

He had never forgotten the photo taken by the Apollo astronaut that hung on the back of his bedroom door – the image of the fragile blue marble, with swirling white clouds, hanging in the dark eternity of space.

He remembered the fear he and his friends had of nuclear war (remember MAD?), of our little planet blowing up in a flash of light, a moment of MAD-ness. Eventually (and thanks in part to children who spoke up), adults had come to their senses, and worked together to make a safer, more peaceful world.

Now Alan knew our little planet was threatened again – this time by climate change – and everyone, especially kids, were worried about it. But this time, he was one of the adults. And he was a scientist, an engineer, and a solver of problems. Alan was determined to help.

One day, he got a letter. It was from his grand-niece, Elise, aged 10.

'Dear Uncle Alan,' it said, 'I have just watched a frighteningly real video about the future of our planet. I would love it if you could talk to my school about what we can do, how we can help and what is actually going on.'

Here's what Alan told Elise and her friends:

Our planet is getting hotter. Elise has already lived through eight of the hottest years ever recorded in Australia. The greenhouse gases we humans have been sending up into the atmosphere from burning fossil fuels like coal and oil – to make the electricity to run our homes, our cars, our planes ... have created a layer, like a blanket, around the Earth. The blanket is trapping the sun's warmth close to Earth instead of letting it bounce off into space. As our planet gets warmer, it's creating more and more problems for the natural world – and for us: rising sea levels as the ice melts, huge storms, droughts, and bushfires.

'But,' Alan reassured Elise, 'the problem was created by technology, so it can be solved by technology'.

Alan believed there had to be better ways to create electricity cleanly, without sending greenhouse gases into the atmosphere. He was already advising on climate change, joining lots of smart people working on lots of good ideas on how we can stop making greenhouse gases. But there was one idea Alan had heard about that he thought might make a big difference. He just had to get everyone to listen. He promised Elise he was going to try.

*

Scientists all over the world have been racing to find clean ways to make electricity. In Australia, our two favourite ways are solar energy (where solar panels soak up the sun's rays and convert them into electricity), and wind energy (the wind turns huge windmills that make electricity for us to use). They are both great ... but not perfect. Sometimes the wind doesn't blow and sometimes the sun doesn't shine (and often at a time when you want to turn on a light).

Solar and wind are heroes, but they can't do it alone. They needed a new hero to join the team and Alan thought he knew just the one: *hydrogen*.

Did you know that the most abundant element in the universe is hydrogen? It's everywhere, even in you. We don't have to mine the bottom of the ocean or dig up our beautiful land to find it. Hydrogen is all around us, including in something you know pretty well: water.

Water is made up of two ingredients: hydrogen and oxygen. If we break up water and use the hydrogen ingredient as a fuel, what's left over? Yep, no nasty greenhouse gases, just beautiful clean water vapour.

Of course, water is precious, but we'd need much less water to make enough hydrogen to make electricity for us than the huge amount of water we use for our mining industry. And we could even make hydrogen with wastewater, **artesian water** or salt water (from the ocean, and there's unlimited litres of that!).

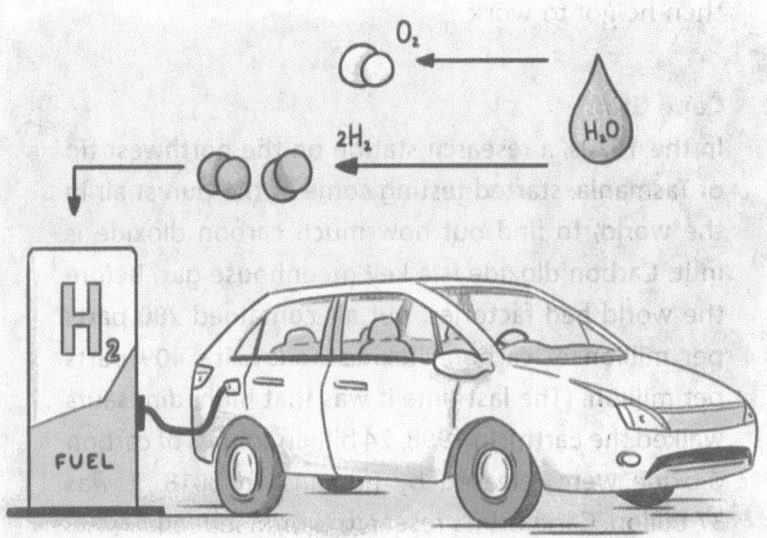

Scientists have already proven hydrogen can light our cities and towns, run our computers, power our factories, heat our homes and drive our cars, trucks and trains. It's clean for our environment, could create thousands of jobs, is safe to use, easy to transport and it doesn't need the sun to shine or the wind to blow.

Best of all, Australia is the perfect place to produce hydrogen.

'What if,' Alan thought, 'we could make enough hydrogen not just to help power Australia, but perhaps the whole world?' He got excited. And then he got to work.

• •

Cape Grim
In the 1970s, a research station on the northwest tip of Tasmania, started testing some of the purest air in the world, to find out how much carbon dioxide is in it. Carbon dioxide is a key greenhouse gas. Before the world had factories, our air contained 280 parts per million of carbon dioxide. Today, it's 409 parts per million. (The last time it was that high, dinosaurs walked the Earth). In 1998, 24 billion tonnes of carbon dioxide were released by humans. In 2018, it was 37 billion. Cape Grim's research is grim indeed.

Jules Verne

Back in 1874, famous science-fiction writer Jules Verne imagined a future where hydrogen would power the world. In his book, *The Mysterious Island*, he wrote, 'water will one day be employed as fuel, that hydrogen and oxygen which constitute it, used singly or together, will furnish an inexhaustible source of heat and light, of an intensity of which coal is not capable'. How amazing is that? That was 145 years ago, and he has inspired scientists like Alan ever since.

Isaac Asimov

Asimov was born in 1919 in Russia and moved to New York as a child with his family. His parents opened a little corner shop that sold comic strip books. The young Isaac loved them, but he became a bit bored by the same old storyline where humans build robots, then everyone dies. Isaac decided to write about a future where humans and robots lived together. How would it work? What would life look like?

So, he became a science-fiction writer and, at the age of 22, he did something truly remarkable. Long before robots or AI existed, he came up with the three laws of robotics that still stand today.

1. A robot must not injure a human being or, through inaction, allow a human being to come to harm.
2. A robot must obey orders given to it by human beings.
3. A robot must protect its own existence.

Who is Ivy?
Ivy is an artificial intelligence robot that was created in Sydney. Unlike doctors, who have to sleep, Ivy watches the little cells that would become babies for 24 hours a day for days and days, without sleeping. She learns from each set of cells she watches, so she can tell doctors which one is most likely to become a healthy baby. Ivy is an example of how artificial intelligence – used well – can help humans.

13

High energy

Bzzz. Bzzz. Alan clutched his phone, fidgeting impatiently.

'Good morning, Alan,' came the bright voice at the other end. Josh Frydenberg didn't sound as sleepy as their last memorable phone call. 'How can I help?'

'Good morning, Josh,' said Alan. 'I was wondering if I could talk to you about hydrogen as a possible clean energy source?'

Josh paused. Alan knew how busy he was. Would he be curious enough to squeeze Alan into his diary?

'Sure, Alan, you did such a brilliant job on the Electricity Review. I'm happy to hear you out on this and learn something new. Make a time with my team. Look forward to seeing you then.'

And just like that, another door swung open.

Before he knew it, Alan was in Josh's office in Melbourne, explaining how hydrogen could help Australia and the planet. Josh had started the meeting a little distracted, but as Alan talked enthusiastically about his vision, he saw a spark of excitement in Josh's eyes.

'I tell you what, Alan,' he said at the end, 'why don't you come to the meeting I have with the other energy ministers and tell them about it.' He paused. 'Perhaps write up a little paper they can take away with them to read, too…'

So, Alan did. But it wasn't a little paper. If he was going to do it, Alan believed in doing it well. He turned up to the meeting with a 60-page glossy booklet full of everything they needed to know. The ministers loved it.

Over many months, dozens of meetings, brainstorming and chats late into the night, that little booklet morphed into Australia's National Hydrogen Strategy. And when the time came for the ministers to vote officially on whether this was something Australia should do, the result was unanimous. Every single minister voted yes. Australia was on its way to being a world leader in clean energy. Alan hoped Elise and her friends would be pleased.

*

By then, four exciting years had rushed by. Alan had often worked seven days a week, usually from before dawn. He was keen for a holiday, but also an adventure, ideally a long way from his desk. Alan, Elizabeth and their sons went to Antarctica on Greg Mortimer's famous ship, to see the world's most pristine continent, and the effects of climate change on it, for themselves.

> **Antarctica** is the driest, coldest, windiest and highest continent in the world. It's nearly twice as big as Australia, and 98% of it is covered in ice. Antarctica contains 80% of the world's freshwater, if it was to melt, the sea level all over the world would rise by 60 metres.

During the trip, Alan was going to use the downtime to plan all the things he wanted to do in his last year as Australia's Chief Scientist. He needn't have bothered. His final year as Chief Scientist was going to be big and crazy – and nothing anyone could have planned for.

> The coldest recorded temperature in Antarctica is **-94.7** degrees C (although it's a bit controversial as it was measured by a satellite in space) which is pretty cold if you think that water usually freezes at 0 degrees!
> Australia is the caretaker of almost half of Antarctica (42%), an area of land that's 80% the size of Australia itself.

The year started with a summer of terrible bushfires. Millions of hectares were burned, more than a billion Australian animals were killed, at least 34 people died and many were left without homes. This was exactly the kind of disaster that scientists, including Alan, had been warning would be made worse by climate change.

The PM called.

'Alan, while you're working on our plans to help with climate change, we also need to be ready for more natural disasters like these awful bushfires.

What could we do better? I need a list. Can you help with that too?'

So, while Alan continued his work on a plan for low emissions technology to help with climate change, he found himself meeting with communities, scientists, and leaders to talk about the bushfires, and make a list of things Australia needed to do to be better equipped for disasters in the future. Ideas like getting satellites that could monitor bushfires from space; manufacturing our own fire-fighting foam here in Australia; finding better ways to get emergency information to people stuck in fires; and organising mental health support for firefighters and anyone caught up in disasters.

Then, while all this was going on, Alan was watching some other news carefully. His fellow scientists all over the world were sounding the alarm. A few weeks earlier, in a city called Wuhan in China, a group of people had come down with a mysterious disease no one had ever seen before. Not only were people there getting very sick, but they were also infecting other people very quickly.

Scientists everywhere were rushing to research it. Leaders were calling emergency meetings about locking down cities and closing borders, and the World Health Organisation (WHO) which is part of the United Nations, met to decide if this new illness they'd named COVID-19 was a global health emergency. It was. And that meant for Australia, too.

Alan's phone started ringing hot.

'Dr Finkel,' drawled a strong American voice, 'this is the White House Office of Science and Technology. We're creating a group of Chief Scientists and advisors from all over the world to work together and share research to defeat COVID-19. Will you join us?'

Of course, the answer was yes.

The Australian Department of Science called: 'Alan, we're worried Australia won't have enough **ventilator machines** to help really sick patients to breathe. Please help!'

Again, yes.

Then it was the Victorian government: 'Er, Alan, we're having a few problems tracing who sick people have been in contact with, so we can warn them to stay home. We need your help.'

And the Australian government: 'Alan, can you join the advisory group looking into the best COVID-19 vaccines and treatments for Australians?

'And um, that contact tracing review you did for Victoria, could you please look into it for the whole country?'

Yes, yes, yes.

*

It was during one of those phone calls that Alan was talking to the PM's Chief of Staff (the leader of the prime minister's team). The conversation went something like this:

'Alan, of course the PM needs good scientific advice, but sometimes the government has to make decisions really quickly, especially during a crisis. Everyone knows that proper scientific advice takes time. Sometimes they just can't wait.'

'John, I know the government is having to make a lot of decisions fast right now. But it's important to have the best possible information first. What if I could get the PM the best advice, from our top experts, written simply, in record time – would that help?'

Alan could hear John's sigh of relief down the phone. He didn't have to say anything more.

And that's how the Rapid Research Information Forum – or 'the RRIF' as Alan nicknamed it – was born. It was a team of Australia's most brilliant minds, from all sorts of specialties, based in our most prestigious academies and universities. They are the quiet heroes of our science world, and Australia needed them all, NOW!

*

Right from the start, the RRIF team was inundated with requests for expert answers that were trustworthy, easy to understand and arrived as quickly as possible. Which experts were called on to help depended on the question. They were often from very different areas: education and medicine;

psychology and physics. Here are some of the questions the government asked:

'If someone catches COVID-19 and recovers, does that mean they won't catch it again?'

'Will our children learn less going to school online instead of being in a classroom?'

'Will winter make COVID-19 spread faster, and will it make people sicker than in summer?'

Alan's RRIF team idea sounds simple, but it worked. It meant the Australian government could ask their RRIF team for any information they needed, and they'd get a detailed answer they could understand, in record time.

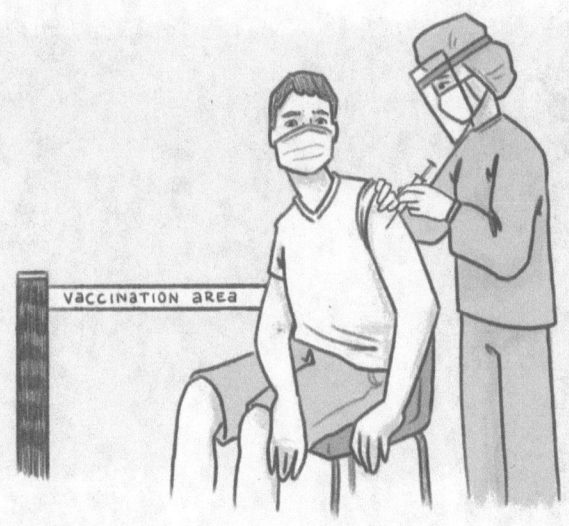

Listening to our brilliant scientists and experts is one of the reasons Australia has managed the COVID-19 pandemic so well – with only 909 people dying from it in the first 12 months while almost every other nation has had thousands of people dying – in the USA, more than 500,000 people died over the same period. The RRIF was an important part of that.

14

What's next?

It was the end of five incredible years, and Alan's term as Chief Scientist. He was a little sad the day he walked out of his Canberra office for the last time. He was proud of everything he and his smart, hard-working team had achieved. He'd miss them and the excitement of being Chief Scientist. But he knew Dr Cathy Foley, Australia's new Chief Scientist, would do a wonderful job.

As he walked through the halls of parliament for the last time, one of the politicians had waved at him, and wished him a happy retirement. They didn't know Alan very well. He wasn't going to try that again.

Driving his much-loved electric Tesla car home to Melbourne, Alan contemplated his new life.

His plan was to split it roughly (and rather unscientifically) into thirds: one-third of his time would be spent on education back at Stile, helping to make science and maths more interactive, more fun and much more interesting for Aussie kids.

One third would be helping the planet. The government had asked Alan to stay on as a Special Advisor for Low Emissions Technology, finding new ways that technology can help with climate change.

And one-third was spending time with Elizabeth, the boys, and of course, his friends. Alan was already planning more adventures for them all.

Most of all, Alan still has those tickets into space, sitting neatly in his desk drawer. Wilson rings Alan regularly to talk about the next possible launch date. Perhaps one day soon the two of them will finally get to blast off on the ultimate science adventure, just as Alan had dreamed about as a teenager who'd skipped school and sat glued to the moon landing, inspired by the possibilities of science and brave endeavour, all those years ago.

> When the government asked Alan to stay on as Special Advisor for Low Emissions Technology, he agreed and offered to do it for free. He felt it was something he could contribute to the world. But it turned out that government contracts won't let anyone put a zero in the money column. So Alan came to a compromise with the government: Australia is now paying Alan Finkel $1 a year to help save the world. And everyone is happy with that.

Afterword

Alan believes the two most important subjects you can study are English and Maths:

English because it's the way we share thoughts and feelings, it's how we reason, and argue, and imagine, and connect. On the practical level, it's how we write the cover letters that get us jobs.

Maths is just as important, because it's the universal language of the modern world. Maths is the language of science, economics, medicine and engineering. It's how we measure the flow of money in business, and the flow of heat in an electric motor. It's how we work out the lifetime cost of buying a home, and how much steel we need so new buildings will not collapse. And both subjects have to be learned, from specialist teachers, in schools.

Alan's passion for maths got a boost in 2018. Australia had an outstanding year in Maths. Our Australian Local Hero of the year was a maths teacher, Eddie Woo. Australian mathematician Geordie Williamson, aged 36, was admitted as the youngest Fellow (or member) of the Royal Society. Then the Australian team won two gold, three silver and a bronze medal at the Maths Olympics. A group of Aussies won the International Mathematical Modelling Challenge for the first time, and an Australian Professor, Nalini Joshi, was elected Vice President of the International Mathematical Union. But perhaps best of all, an Australian won the Fields medal! (It's like the Nobel Prize for Mathematics, but it's only announced once every four years, and only 60 people in history have ever won it). It was awarded to Professor Akshay Venkatesh, also aged 36, from Perth.

Glossary

- **Amplifier:** An electrical device that increase the power of a signal.

- **Artesian water:** Water from a natural underground reservoir. Australia has the largest and deepest artesian basin in the world: the Great Artesian Basin, stretching over 1,700,000 square kilometres. It is the only source of fresh water for humans and farms through much of inland Australia where there is very low rainfall each year. The water is brought up from underground with bores that are drilled down to where the water is.

- **Biochemistry:** The study of chemistry of a living body.

- **Cystic fibrosis:** A serious diseas that causes lungs and other organs like the liver and pancreas not to work properly.

- **Electric current:** A stream of charged particles, electrons or ions, moving through an electrical wire or a space.

- **Electric voltage:** The force of an electric current, measured in volts.

- **Electrode:** The point at which an electric current enters or leaves something, eg. a battery.

- **Entrepreneurs:** People who start their own business, especially when this involves seeing a new opportunity or innovation/invention.
- **Epilepsy:** A condition of the brain that causes a person to become unconscious for short periods or to move in a violent and uncontrolled way.
- **Metallurgical engineering:** Working on converting raw metals into useable materials, recycling metals into new products, etc.
- **Neuroscience:** The study of the nervous system and the brain.
- **Nobel Prize:** Five separate prizes are awarded in the fields of Physics, Chemistry, Physiology or Medicine, Literature, and Peace, to 'those who, during the preceding year, have conferred the greatest benefit to mankind' (Alfred Nobel).
- **Parkinson's disease:** A disease of the nervous system that makes the muscles become stiff and the body shake and gets worse as the person gets older.
- **Pneumonia:** A serious illness where the lungs become red and swollen and fill with liquid that makes it hard to breathe.
- **PhD:** It stands for Doctor of Philosophy but is a research degree that takes about 4 years of full-time to complete after completing a Bachelor's degree and an Honours year or a Masters in the subject area.

- **Postdoctoral research:** Advanced work or study that someone does after they complete their PhD.

- **Quantum physics:** Studies the behaviour of matter and energy at the tiniest microscopic levels.

- **Senate hearing:** Senators belong to committees that examine particular issues the senate is considering. Experts or relevant people are asked to present in person or in writing to them.

- **Silicon Valley:** A region in California that developed as a global centre of technological innovation.

- **Ventilator machines:** Help people breathe when they have difficulty breathing on their own, by forcing air in and out of their lungs.

About Kim Doherty

Kim Doherty is an editor, storyteller, teacher, and a mum to two young children – Molly and Xavier – who she hopes will be inspired by the amazing world of science and Alan's story. This is Kim's second book for children, the first was about Mt Everest and the amazing facts, death-defying adventures and strange myths of the world's highest mountain.

When she's acting like a grown-up (which is not very often), Kim is an award-winning editor-in-chief who has led some of Australia's best-loved publications, from *The Australian Women's Weekly* to *Kidspot*. She is passionate about sharing Australian stories and the importance of lifelong learning: Kim is hoping to complete her most recent educational endeavour (the world's slowest Master of Laws) sometime before the end of the century.

Dedication

For Molly, Xavier and Ben.

In memory of Vera Finkel who died during the writing of this book.

Many thanks to Elizabeth Finkel, Vivienne Serry, Ron Finkel, Amanda Caldwell, Anne Marie Lansdown, Wilson da Silva, Rita Nash and the Sydney Jewish Museum for their generous assistance with research.

Grateful thanks to Ben Doherty and Catherine Lewis for their patience, support and editing brilliance. And Diana Silkina for her illustrations.

Special thanks to the inspiring Alan Finkel for his patience, expertise and enthusiasm – and for kindly finding the time to share his memories and wisdom with the children of Australia.

Aussie STEM Stars

If you loved this story and want to read about more of our great Aussie STEM Stars, check out: AussieSTEMStars.com.au to see what's available now as well as some of the upcoming stories of fabulous Aussies changing our world!

You can buy them at your favourite bookshop or from AussieSTEMStars.com.au

AVAILABLE NOW

A plastic surgeon, **Fiona Wood** revolutionised treatment for burns victims, improving hundreds of lives with her spray-on skin invention. She came to world-wide fame treating the Bali bombing burns victims. Her story shows us the value of dreams, hard work, and the courage to do what's right.

ISBN: 9781925893281

Eddie Woo has been an Australian high school Maths teacher, education ambassador and advisor, author, TV Host and 'WooTube' sensation, and named as one of the World's Top 10 Teachers. He's a champion of school-based integrated STEM education.

ISBN: 9781925893403

Georgia invented a taste-aversion technique to prevent native fauna from being decimated by the rampaging toxic cane toads in northern Australia. She has implemented many community conservation programs, including collaboration with indigenous ranger networks.

ISBN: 9781925893342

As a surgeon, **Munjed** fled Saddam Hussein's regime in Iraq after refusing to cut off the ears of army deserters. Fleeing to Australia by boat, he was locked up in the notorious Curtain Detention Centre. He is now the world leader in osseointegration surgery and 'clever' prosthetics.

ISBN: 9781925893373

Creswell Eastman AO is the Clinical Professor of Medicine at Sydney University Medical School. He has directed public health projects into elimination of iodine deficiency disorders around the world.

ISBN: 9781925893526

Gisela Kaplan's story begins in post-war Germany. She endured many challenges that led to a profound curiosity, care and compassion for all living beings. Her scientific studies on some of Australia's iconic bird species, have helped her become a leading voice in animal behaviour worldwide.

ISBN: 9781925893465

When **John Long** clamped his hands around a fossil at the age of seven, it sparked a lifelong love affair with collecting artefacts from another time. Currently serving as the President of The Society of Vertebrate Palaeontology he has still found time to write 26 adult and children's books about fossils.

ISBN: 9781925893687

COMING IN 2022

Marita Cheng is a robotics entrepreneur and the founder of the global not-for-profit organisation RoboGals. She builds robots that help people with disabilities to live fuller lives. She's the youngest member of the Order of Australia, and the 2012 Young Australian of the Year.

ISBN: 9781925893557

Underwater explorer, **Emma Johnston** comes from a long line of seafarers and grew up beside the ocean. So, it's hardly surprising that she's now a world leading marine biologist. Emma's an inspiring advocate for women in science and has received many awards and an Order of Australia for her vital work.

ISBN: 9781925893558

 **Michelle Simmons,** 2018 Australian of the Year, is a pioneer in atomic electronics and quantum computing. Currently Scientia Professor of Physics at the University of New South Wales, she strongly encourages girls to pursue careers in science and technology.

ISBN: 9781925893496

AND MORE TO COME ....

Keep up with our latest news and releases at:
www.AussieSTEMStars.com.au